About the Author

Richard Keegan, BE, CEng, works in the Manufacturing Consultancy Service of Forbairt, the State Development Agency for Indigenous Companies in Ireland. He qualified as an engineer from University College Dublin in 1980 and has worked in both engineering and marketing capacities for major domestic and multinational companies such as Nokia and Nestlé. His work for Forbairt has been largely in the SME sector where he has pioneered the introduction of World Class Manufacturing principles in Ireland. He is currently leading a project on behalf of the European Commission on Benchmarking for Competitiveness — Improving World Class Performance.

An Introduction to World Class Manufacturing

Richard Keegan

Daniel N Kvy

Oak Tree Press
Dublin • London

Oak Tree Press
Merrion Building
Lower Merrion Street
Dublin 2, Ireland

A previous edition of this book was published in 1995 as
World Class Manufacturing . . . in an Irish Context,
co-authored with John J. Lynch

A catalogue record of this book is
available from the British Library

ISBN 1-86076-052-X

Printed in Ireland by Colour Books Ltd.

Contents

Acknowledgements

This book is based on people. The ones who have the ideas and the ones who make them a reality. The ideas people get enough attention. We would like to thank some of those who are helping to make the ideas a reality:

Malcolm Lewis, Dave Neacy and their team; Martin Shiel, John O'Neill, Bob Galavan and team; Liam Lacey, Lee Choo, David Thackerberry and team; Phonsie McEntaggert, John McEntaggert, John McMahon, Larry, Ricky and team; Jackie O'Dowd, Terry McMahon, Olive, Helen, Edel and team; Don Curry, Donal and team; Louis Mulcahy, Peter Wall, Steve Ó Culáin, Donal, Maeve, Ricky and their team.

Many other companies are actively pursuing WCM principles. We wish them luck and perseverance. My sincere thanks to Dr John J. Lynch for his general support and specific contribution to the strategic chapter of this book. Special thanks to Agnes, Paula and David Givens for their patience, Denis for his uncommon sense, and Helen for always seeing more than one side of a problem.

Foreword

In the last decade it has been my enormous privilege to visit and work with factories in almost every corner of the globe. I have walked the production lines with Chief Executives and operators alike. I have my favourite factories as well as those I am in no hurry to revisit. Factories are my "parish". When I reflect on this experience I can begin to dream of a manufacturing company that exhibits all the very best practices I have seen anywhere in my travels, a factory that is "world class". The dream goes something like this:

The inspirational leadership of the Chief Executive provides a clear vision for the direction of the business. The strategy is jointly developed and shared throughout the site at every level. Employees are inspired to follow the direction set and are encouraged and trained to take responsibility for its achievement. Everyone is involved and feels valued and is keen to offer to undertake multiple roles. The measurement of the business (as seen through the customer's eyes) is displayed for all to see. Customers are encouraged to be regular visitors to the site. It is the operators themselves who show customers around.

Relationships with suppliers are built on the assumption of lasting partnership. The basis is one of trust rather than mistrust. The benefits of joint efforts to reduce total supply chain costs are shared with the supplier. Outbound logistics are capable of delivery into highly variable demand patterns of Just-in-Time retailers, for example.

Information technology systems are integrated so that the design process can, for example, deliver a workable bill of materials to the planning process. CAD and CAM can realistically be spoken of in the same breath. MRP schedules are trusted and acted upon without need for local modification.

> *Every aspect of the manufacturing processes that adds cost but
> not value has been systematically eliminated (unnecessary
> movement, counting, inspection, paperwork, etc.). The design
> and development process involves suppliers and customers as
> well as manufacturing and sales teams. The product will not
> only meet customer requirements, but will enable optimum
> manufacture, distribution, ease of use and end-of-life disposal.
> With all business processes sharply focused on meeting and ex-
> ceeding customer expectations, a spirit of continuous improve-
> ment pervades the entire business.*

It has become something of a joke amongst my colleagues that
before you look at the bookshelf in a manufacturing executive's
office, it is possible to predict the books they will have with some
degree of certainty. There is apparently no great secret about the
World Class Manufacturing vision. What we must then ask is why
so few manufacturing sites come close to fulfilling the vision.
What is worse is that so many executive teams believe their factories are much closer to this role model than is the case. I am
always being told that the journey to World Class is three years
away, even by factories whose progress appears slower than the
pack, for whom a more realistic assessment is that, at their current rate of progress, they will never catch up.

The answer is, at least in part, that we can all talk about the
vision, yet too few can fill in the practical detail behind the concepts. Much of the Japanese-influenced best practice in Lean Production and Total Quality is firmly rooted in practical common
sense. We know, however, what a scarce resource that is! Too often, I am told that Just-in-Time doesn't apply in this industry or
that Total Quality Management didn't last long enough to make a
difference. There is nothing wrong with Lean Production or TQM,
of course; it is the implementation that has failed in these cases.

I am delighted that Richard Keegan has taken the time to
bring his experience together in this very practical book. What he
has achieved is to produce a single text that catalogues the practical detail and common sense behind the vision. I believe it will
be of equal value to engineers and accountants and to large and
small sites. My travels suggest that the very practical approach

adopted in this book will be of universal value to anyone running a manufacturing operation, anywhere.

My great hope is that this becomes one of those books that is read and re-read, used for reference, loaned out and given away. Its measure of success will be that it is not stored away as an ornament on the Chief Executive's office bookshelf, but lives in the everyday life of the cell leader's team drawer. I hope this book will be "dog-eared" before it is discarded, as one day it will, as the moving target of what is needed to become and remain "world class" moves ever onwards.

Philip Hanson,
Principal,
IBM Consulting Group.

Chapter 1

Introduction

World Class Manufacturing (WCM) is about competitiveness. It is the name given to a modern revolution taking place in the most competitive manufacturing operations in the world today. World Class Manufacturing can be difficult to define, but like quality or electricity, we all know it when we see it!

One of the problems in defining WCM is that it is known by different names in different countries. Lean Production, Value Management, Business Process Re-engineering, and Total Quality Management are each used in Europe to describe WCM-type processes. At a recent international gathering, however, the general consensus was that the common thread between all of these new processes was that companies embracing them achieved major competitive advantages in terms of quality, cost and lead time.

This book will not try to give a definitive answer to this question of what exactly is WCM; rather it will give a guide to understanding the principles and, more importantly, some practical elements of World Class Manufacturing that can be implemented in your particular company. For the purposes of this book we will define WCM as follows:

World Class Manufacturing means making products

Quicker
Better
Cheaper . . . Together

That is the essence of World Class Manufacturing. A more wordy description, however, would be:

World Class Manufacturing is the pursuit of superior performance in Quality, Lead Time, Cost and Customer Service through Continuous Improvement in Just-in-Time (JIT) Manufacturing, Total Quality Management (TQM) and Employee Involvement.

World Class companies optimise the problem-solving abilities of their employees in applying both modern techniques, such as JIT and TQM, and traditional engineering processes. This book will develop these ideas and present a number of simple-to-apply techniques that will assist you in your efforts to implement WCM in your own company. The emphasis will be on simplicity — simplicity of techniques as well as implementation methods. This thrust towards simplicity is the true essence of World Class Manufacturing.

WORLD CLASS MANUFACTURING — WHERE DID IT COME FROM ?

The past 20 years have seen many changes in our personal and business lives. In particular, competitive pressures in the global economy have meant drastic changes in the way businesses are organised.

In the 1970s, Japanese companies started to appear on the world markets as serious competitors, across many different market sectors, and the once all-powerful British and American economies began to take a battering. While Japanese products had been belittled in the 1960s they had become highly respected and sought after a decade later. Through the 1980s the Japanese built on their successes and came to positions of superiority in many areas, including, cars, motorcycles, hi-fi equipment, cameras, video and CDs.

This major shift in market position started to ring alarm bells in Western management circles. In America, Harley Davidson asked for Government protection. In Europe, quotas for car imports were agreed with the Japanese to protect the European manufacturers. Japanese firms responded by setting up manufacturing plants of their own inside the "fortresses". They thus transplanted their manufacturing systems to America and Europe.

Western researchers now had the opportunity to see, at first hand, just what it was the Japanese were doing to gain such superior market positions. As the transplanted companies were located in Western countries their efficiency could not be explained away by cultural differences. What the researchers found was that the transplanted companies were using such tools as Just-in-Time Manufacturing, Total Quality Management and Employee Involvement — new processes which were to revolutionise manufacturing around the world.

FIGURE 1.1: EVOLUTION OF WCM

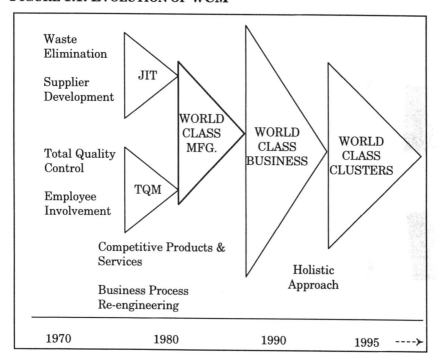

In the mid-1980s an American researcher, Dr Richard Schonberger, visited Japanese companies and Japanese transplants in America. He coined the phrase World Class Manufacturing to encompass what he had observed happening in the best of the Japanese firms, where Total Quality Management, Employee Involvement and Just-in-Time techniques were being applied.

Since the mid-1980s, the base concept of World Class Manufacturing has become fully developed and has now evolved to the World Class Business stage. And now the World Class Cluster concept is being put forward to describe a number of companies aligned along the value chain from first level supplier through to the final user. This concept will be dealt with later in the book.

So what are Just-in-Time, Total Quality Management and Employee Involvement?

Just-in-Time Manufacturing

Toyota and other Japanese firms created a system for manufacturing based on the Toyota Manufacturing System which they called Just-in-Time Manufacturing (JIT). JIT, as its name suggests, means making parts or providing a service as it is needed — not earlier and certainly not later.

To produce "as required" puts a lot of strain on a manufacturing operation — workers, machines and materials all have to be just right, capable of performing at their best. Co-ordination between manufacturers and sub-suppliers also needs to be very good. The major Japanese manufacturers developed a number of tools and techniques as they perfected their own versions of JIT, including:

- KANBAN — a visible control system

- JIDOKA — a process for adding low cost, stop/start and warning systems to machines

- ANDON — an early warning, "trouble lights" system

- Single Minute Exchange of Dies

- Layout and Flow Analysis

- Set-Up Time Reduction

- Manufacturing Control Systems

The Japanese version of JIT reflects their own culture and national characteristics. For example, because of limited space and difficult infrastructural conditions, they needed to find a way to manufacture without stock piling up. Western companies are now looking at these concepts to find an interpretation that suits their cultures and specific needs.

Just-in-Time Manufacturing is discussed in more detail in Chapter 2.

Total Quality Management

In the 1950s Edward Deming, a renowned American statistician, and other researchers visited Japan and introduced them to the possibilities of product improvement using quality tools. The Japanese quickly understood the importance of quality, and Japanese product quality went from poor to superior in less than 15 years.

The Total Quality Management (TQM) movement started the process of examination across a company's value chain. TQM addresses the company-wide quality aspects of a World Class Business where the efforts of employees are focused on "doing things right — the first time", on developing products and processes to the point where mistakes are impossible, and on continually striving to provide ever better products or services.

Total quality in a World Class company differs from the more traditional "measure and test"-type quality system, which can be regarded as passive or at best reactive, as faults are counted and reported upon, and little more. In the WCM company — where the quality ideal is all-pervasive — the quality system is active, and in the better companies, aggressive, in its efforts to identify and eradicate problems and develop processes.

The four main pioneers in the field would be Dr W.E. Deming, Dr J.M. Juran, Dr K. Ishikawa and Mr Phil Crosby. While each would put their own slant on the subject, there is very much greater agreement than disagreement between them. Dr Deming's 14 points for management in terms of quality are as follows:

1. Create constancy of purpose towards improvement of product and service.
2. Adopt the new philosophy. We can no longer live with commonly accepted levels of delays, mistakes, defective workmanship.
3. Cease dependence on mass inspection. Require, instead, statistical evidence that quality is built in.
4. End the practice of awarding business on the basis of price tag.
5. Find problems. It is management's job to work continually on the system.

6. Institute modern methods of training on the job.

7. Institute modern methods of supervision of production workers. The responsibility of supervisors must be changed from numbers to quality.

8. Drive out fear, so that everyone may work effectively for the company.

9. Break down barriers between departments.

10. Eliminate numerical goals, posters and slogans for the work force, asking for new levels of productivity without providing methods.

11. Eliminate work standards that prescribe numerical quotas.

12. Remove barriers that stand between the hourly worker and their right to pride of workmanship.

13. Institute a vigorous programme of education and retraining.

14. Create a structure in top management that will push every day on the above points.

A wide range of quality tools have been created and used over the years to achieve the above goals. Chapter 3 on TQM will address the most frequently used tools, with examples of how they can best be applied in small and medium-sized companies.

Employee Involvement

The third pillar of a World Class Manufacturing company is its people. In this context, the idea of Employee Involvement is crucial. In the old-style company, the work was broken down to the simplest possible element. Henry Ford pioneered this idea to the point where he could bring in new staff and with the absolute minimum of training have them working productively within hours — doing a very simple job. The world has changed. Now, a company may well have to produce many variants of its products, not just a single, black Model T! By employing the latent abilities and skills of its workforce to address and help develop its products and processes, a World Class Company can make significant improvements in both.

Employee involvement allows a company to apply the thinking of *all* its staff to address its problems, rather than solely relying on "management" to be all-knowing. Chapter 4, Employee Involvement, will address the tools and techniques of Employee Involvement in more detail.

WORLD CLASS MANUFACTURING AND COMPANY STRATEGY

Company strategy is a complex subject with many volumes dedicated completely to it. Basically, strategy is the term used to identify the future direction for a company and how management expects to achieve the vision it has formulated. Overall, company strategy will typically examine where a company is going, what markets it should be in, what product areas it should be providing to its customers and so on.

World Class Manufacturing relates to the *operational* side of strategy. It focuses on how the company can manage its basic systems and processes to a very high level in order to achieve the strategic goals set out by top management. When the operational side of the company has achieved an effective understanding and practice of the World Class principles it is very capable of shifting its abilities to react quickly to any changes in the company's overall mission. A World Class company can support the strategy of its top management, because it is World Class.

For a fuller discussion of company strategy and World Class Manufacturing, see Chapter 9.

HOW TO APPLY WCM

The most important aspect of any idea is what you do with it. WCM is essentially a new way of looking at your business. The benefit of WCM to you is how you implement it within your own company and what benefits accrue from this. The five basic steps needed to implement a World Class Manufacturing programme are listed below, and Figure 1.2 shows the amount of change associated with each step.

FIGURE 1.2: STEPS IN WCM PROGRAMME

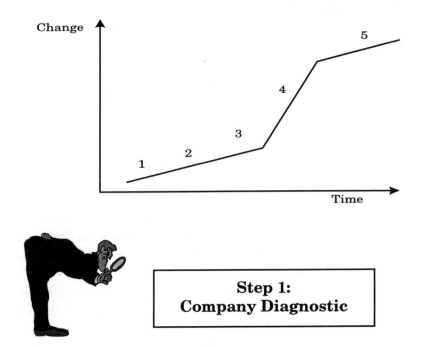

Before moving to a "new land" it is advisable to know where you are starting from. The company diagnostic is effectively a company-wide review. It should look at the key areas of the business, examining Finance, Marketing, Selling, Materials, R&D and, in particular, Operations. A benefit of this process can be the identification of basic problems within the business — problems that can be addressed and solved simply. Quite often, a cold-blooded examination of the existing business can have a very positive effect on the bottom line.

This diagnostic phase is best carried out by an external consultant. This person will generally come to your company with a new mind and fresh eyes, unhindered by the day-to-day experiences of the business. Working with this person can be very worthwhile for the company.

Step 2: WCM Awareness and Self-Assessment

WCM programmes are generally driven by the senior managers of a business, which means that the senior managers must have a clear understanding of what WCM involves in order to pass on the word to the rest of the company.

The key individuals involved need to understand the main principles of WCM and, more importantly, appreciate how those principles can best be put to use in their own company. Once the company has come to an understanding of the basic principles it can compare the results of the diagnostic phase against these principles as a form of self-assessment. This assessment will be used as a basis for the implementation plan for the company.

This interpretation process of the basics of WCM will lead to the planning and implementation phase.

Step 3: Implementation Planning

The company has now learned about WCM and interpreted these principles for its own situation. The company diagnostic phase has identified a number of potential areas for improvement. The combination of these two elements allow the company to move to create an implementation plan.

A successful implementation programme needs to be both practical and flexible. Practical improvements achieved at an early stage in an implementation programme will have a very positive effect on the morale of all concerned.

The plan should be balanced between detail and flexibility. It is generally better to err on the side of flexibility at the planning stage as this leaves the way open to getting the best out of the full team both at management and worker levels, as the process continues.

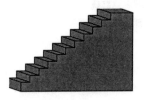

Step 4: Step Change

The examination has now taken place; the company has learned about and come to its own understanding of what WCM is. The planning has also taken place, and now it is time to implement the changes.

One of the features of WCM is that at this point in the process relatively major changes begin taking place within the company. This generally happens as a concerted push as the operation moves into a new gear.

This stage is generally the most exciting and interesting part of a WCM programme as employees are seeing major change and generally major improvement. This is also the most dangerous part of the process. Because the company is moving through a process of serious change, the staff's reaction and perception need to be managed. It is at this point that the importance of the previous two steps — WCM awareness and implementation planning — really comes home to the company.

Step 5: Continuous Improvement

The initial stages of a WCM programme will lead to a number of immediate improvements in the general operation of the company. It is important to remember, however, that the company will need to continue the improvement process into the future as well.

It is strongly recommended that the company prepare a plan at the outset of the programme to provide for training and ongoing assessment to ensure that the WCM principles become ingrained in the company into the future.

The above is a brief description of a WCM programme. The following chapters will provide a detailed look at some of the specific techniques that have been found useful and applicable in numerous small and medium-sized companies throughout the world.

Chapter 2

Just-in-Time Manufacturing

Though the Japanese have been credited with the creation and development of the idea of Just-in-Time Manufacturing (JIT), in reality it has been around now for many years. The typical example used to explain JIT is that of a boat, sailing over submerged rocks, as shown in Figure 2.1.

FIGURE 2.1: SEA OF STOCK — BEFORE WCM

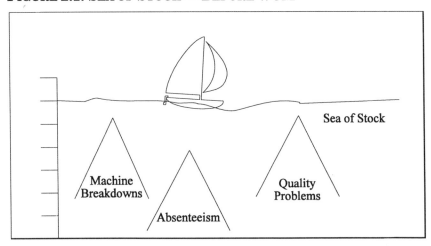

The sea of stock has traditionally kept the boat afloat. However, stocks cost money. It costs money to buy the raw materials, manufacture the goods, pay the workers and store the finished goods until the customer finally wants them — one hopes. Sometimes we even end up with obsolete stocks which are either dumped at a total loss, kept on a balance sheet or sold at heavily discounted rates to recoup some of the cost. These sales can have an adverse effect on the market place, lowering the market perception of your products and keeping market prices low.

Figure 2.2 illustrates how the application of World Class Manufacturing concepts and tools reduces machine breakdowns, quality problems and absenteeism. By addressing these problems, the amounts of stock held — both as finished goods and work-in-progress — reduces the basic working capital requirements of the company.

FIGURE 2.2: SEA OF STOCK — AFTER WCM

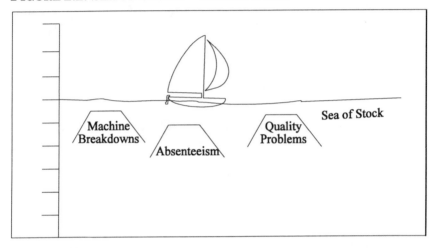

JIT focuses on waste. Not waste as we understand it in a Western sense, but waste in a true sense — any cost added to a product, or service, that does not add value to it. By actively and aggressively addressing the causes of waste within an operation we can reduce the costs associated with a product. This provides us with an opportunity — an opportunity to compete more effectively in the market. Production effectiveness can be a powerful weapon in a company's overall strategy. The marketing department can now decide to increase profitability or to go for increased market share as the company will be operating from a reduced unit cost base.

Toyota have been one of the leaders in these areas of JIT and waste identification. Jujio Cho of Toyota, one of the pioneers in the field, defines "waste" as "anything other than the minimum amount of equipment, materials, parts, space and workers' time, which are absolutely essential to add value to the product".

In a normal factory we usually think our operations are well organised, efficient and well managed. We tend to rely on supervisors to control the operation, to track products and orders, to expedite "rush orders", to count production and "watch" the workers. Unfortunately, built into many of our operations are various kinds of waste. Some of the typical wastes include:

- Carrying heavy pieces
- Double handling
- Counting
- Over-production
- Watching the machine run
- Moving parts a long way
- Looking for parts or tools
- Machine breakdowns
- Lack of parts.

A very simple exercise can be carried out in practically any industry, be it manufacturing or service. Examine the difference between the amount of time it actually takes to process a part, provide a service, or serve a customer and the total amount of time from receipt of order to completion. In a typical factory we find:

- Lead Time — the total amount of time from receipt of an order to delivery to the customer — is 240 hours (six weeks).

- Processing Time — the amount of time actually spent working on the product — is 40 hours (one week).

Table 2.1 gives indicative levels of lead time and processing time for some typical industries.

TABLE 2.1: PROCESSING TIME VS. LEAD TIME IN VARIOUS INDUSTRIES

	Electronics	*Heavy Engineering*	*Light Engineering*	*Medical*
Processing Time	4 hours	40 hours	8.5 hours	56 hours
Lead Time	160 hours	240 hours	56 hours	368 hours

The differences between Processing and Lead Times is an indicator of the level of waste in the production environment — and the old adage that "time is money" is certainly applicable in manufacturing. This waste can be broken down under the three main headings of "people", "materials" and "machines", and can be represented simply as shown in Figure 2.3. In this diagram, the productive work is represented by the Processing Time, and the difference between Processing Time and Lead Time represents the waste.

FIGURE 2.3: WASTE IN FACTORIES

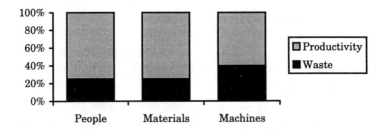

The full thrust of JIT is in identifying the wastes and then eliminating them completely, where possible, or at the very least reducing them significantly.

Management gurus in the United States and Europe have studied this question of waste for some years now. Initially they believed that over 60 per cent of waste was caused by the systems workers were asked to follow, but over the years this value has increased to 80 to 90 per cent of total waste being attributed to the systems, methods and processes that management tell their workers to use! A nice point about this research is that these wastes are possibly the easiest to address. It means, however, that managers must open their collective minds to accept that they are responsible for the solution as well as the problem.

THE SEVEN WASTES

Now let's return to Toyota to the question of what makes up the waste content of our systems of production or administration. After many years of development work, Toyota came to define seven types of waste:

1. Waste from overproduction

2. Waste from waiting time

3. Transportation waste

4. Processing waste

5. Inventory waste

6. Waste of motion

7. Waste from product defects.

Waste from Overproduction

What can be more wasteful than making more than you need? The question is rhetorical. However, a walk through many companies will show people working busily, generally making for stock, hoping the customers will actually want to buy that particular product, that particular size or colour. In an expanding market companies can often get away with this, but markets aren't always expanding and all too frequently companies have to be able to react to downturns.

The focus on production needs to be shifted to producing only what is required by the next step in the process. In actual fact, this means looking at the last step in the process first and working back down the line. This regimen puts severe pressure on the selling and marketing people to get close to their customers and markets to allow them to provide accurate forecasts to the operational people to allow them to manage the flow through production to meet customer demands.

Getting a production operation to the point where it can react effectively and efficiently to changing market demands requires inter-disciplinary working and understanding. The use of KANBAN (demand control system), quick changeover techniques, cellular manufacture and the wide range of other JIT techniques can be a great benefit in achieving this.

Waste of Waiting Time

Waiting time is usually easy to see but hard to recognise. For example, most people accept that a machine "needs to be watched", but, in reality, with most modern machinery this is no longer the case. Usually machines nowadays have sophisticated control systems that often allow them to run either automatically or semi-automatically. Even older machines can often be easily equipped with early warning devices — lights/bells or buzzers to alert operators if a material supply is about to run out or if a finished goods container is nearly filled.

If a person is "machine watching" because the machine is prone to failure or frequently produces bad product, then the machine watcher is not actually solving the problem. There is something fundamentally wrong with this machine that needs to be engineered out of it. The longer the machine is watched, the longer the problem will persist. It is much more efficient to fix the root cause of the problem than to constantly address the symptom.

Transportation Waste

Double and treble handling of parts, components and finished goods is a widespread practice. For example, quite often goods are part worked, put into a tray or moved to a work-in-progress (WIP) area, or temporary store. Labels or some means of identification have to be written for the goods to satisfy ISO or BS requirements. When the next process stage needs the goods an "in process co-ordinator" as some multinationals call them, or stock handler in an indigenous company, has to go and locate the particular bins, boxes or trays of goods and move them to the production area. Generally a supervisor will decide that the goods are required, look for them, look for the stock mover and tell him what he wants moved, where and by when. There should be no need to discuss any further the pure wastefulness of this process.

In an even worse scenario, if a factory has been laid out along functional or departmental lines, the distances to be travelled will be long and generally require the use of a pallet truck or fork lift truck. Recent work in one manufacturer showed an average travel distance of parts of 1,747 yards! Layout improvements, control of production and general organisation can have a very positive effect on transportation wastes.

Processing Waste

Probably the first and most important question that can be asked when moving towards a WCM implementation in a company is: "What are we trying to do?" This sounds like a simplistic question but like all such questions it can have a monumental impact on the person of whom it is asked.

Unfortunately, the usual focus for most companies when they start to look for performance improvement is: "How can we do what we do, better?" The answers to this question usually result in incremental improvements of about 5 to 10 per cent in overall efficiency. This once would have been quite acceptable to traditional Western management, but today's companies are competing in an increasingly international market. The Pacific Rim nations, for example, are achieving improvements of 30 to 60 per cent against our 5 to 10 per cent improvements.

To achieve the same level of improvements as our international competitors we have to fundamentally change our way of looking at our processes. We have to start by asking the question: "What are we trying to do?" When applied to our processes, the answer can and should lead to major improvements.

As an example, say a small company manufactures an item with a processing time of 36 days. They have come under pressure from a Japanese competitor in the market place. The Japanese company appears able to supply customer requirements immediately without recourse to large stocks of finished goods. Customers now expect a turnaround of 4 weeks from order placement to receipt of goods. This is equivalent to 20 working days. Clearly the small company has a problem. With a production cycle of 36 days it can only hope to meet the customers' expectations by holding large and costly stocks of a wide range of products. The company's research indicates that the Japanese company has a production cycle of 9 days, allowing them two full cycles of production within the customers' expectation window, without recourse to stock holdings.

A careful examination of the small company's production process showed that a large bloc of time was allocated to a specific sterilisation process. In the original process, the products were sterilised in bulk outer cartons, which resulted in a slow bleeding-off of the sterilising gas before safe levels were reached. This resulted in a sterilising time of 35 days before the gas concentration reached a safe level. In the suggested process, the items are sterilised individually on an exposed surface, which allows the gas to bleed off much more quickly resulting in a safe level being reached in only 8 days. This lateral approach to processing waste can provide the potential for significant improvements in processing times.

The same logic can be applied to a mechanical assembly operation. For example, assume a part needs to be drilled to allow for a bolt to be pushed through to join two pieces together. This machining operation can be done by using a hand drill, a drilling machine right up to an automatic loading/unloading machining centre. Each of these evolutionary steps can reduce the processing time. But what happens to the processing time if the part is redesigned to be assembled as a snap-fit?

Inventory Waste

Inventory waste is tied closely with over-production waste. It requires space, people and money to hold it. The "benefit" of inventory is that parts are generally available to allow for smooth production — in theory at least. In practice the problems involved with inventory management generally outweigh the advantages. More often than not, production in a traditionally-managed plant will be held up for lack of often the simplest and cheapest part. Unfortunately, for detailed Materials Requirement Planning (MRP) systems to have a chance of operating effectively, stock accuracy has to be in the high 90 percentile, which can be difficult to achieve.

Probably the biggest reason to attack inventory waste, however, has nothing to do with the inventory itself. The availability of inventory can allow for the shrouding of other problems within the factory. When stocks of parts are reduced, on the other hand, it becomes ever more important that all parts produced are good quality, that production machinery operates efficiently and consistently and that staff are present and available for work when they are needed. In fact, it is often when inventory levels drop that other problem areas can be identified and resolved.

Waste of Motion

People like to be busy, but unfortunately being busy does not always mean adding value to a product. For example, a worker can spend a lot of time each day looking for tools or parts. All that time is waste.

Parts, components and tools should be arranged to minimise the amount of effort required of a worker. If an operator is to run several machines, the machines themselves should be located to minimise this effort. Managers should look at their own environments and see how often somebody is "busy" without adding a penny worth of value to the product or service they are selling.

Waste from Product Defects

If one machine produces a defect it slows down the next step in the process. This is patently obvious. In a traditional manufacturing operation a machine produces parts, which are put into WIP for a period of time. When the next processing step comes to use the parts they may be OK, but quite often they are not. This leads to rework and lost production output. By reducing the lead times in an operation and improving the interaction of production processes, managers can reduce the incidences of product defects.

A number of industrial engineering techniques can be of great benefit in reducing the incidence of product defects.

1. *Identify the root cause of problems*. Don't just accept a machine as being problematic, identify what is actually giving trouble and totally and completely put it right. Many factories operate an extremely good "fire-fighting service", but these can be very costly and generally ineffective in the overall scheme of things. When you "put out" a fire, leave the area so wet that no other fire can be started there again.

2. *Develop the Process*. It is possible to simplify the process so that individual steps can be completely eliminated. Make the

process as simple as possible. Quite often simple tools or jigs can be developed to make it impossible for the part to be processed wrongly. Basic levels of automation, either of the process itself or of the early warning system, can avoid later difficulties. The simple warning of a light to indicate raw material levels are low, for example, can often stop a machine jamming or stopping.

3. *Train the staff.* Make sure your people know how to do what needs to be done! It is commonly accepted that the people actually doing the work know best how it should be done. But only if they've been trained well. Indeed, they may know how to do what they are doing, but this may not be the best way of doing what needs to be done! In other words, the basic process needs to be examined and developed critically. Team workers' inputs can often be invaluable if a process is to be developed. Listening to workers with an open mind can often result in simple, cheap and cost-effective solutions to problems you didn't even know you had!

PROCESS AND PHYSICAL FLOW

Many people will have had exposure to the idea of process flow. Generally an integral part of all quality systems, process flow lists the specific stages in a process, as shown in Figure 2.4.

FIGURE 2.4: ELECTRONICS FIRM PROCESS FLOW DIAGRAM

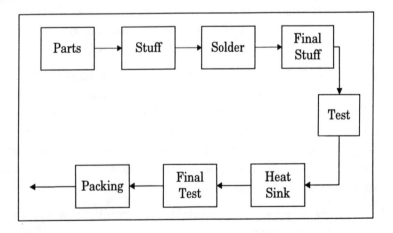

Production stages typically take place in particular departments under specific supervisors. These supervisors concentrate on their own department's performance and watch their own people and processes. The following two examples of distinctly different industries show this type of arrangement.

FIGURE 2.5: ELECTRONICS FIRM PHYSICAL LAYOUT

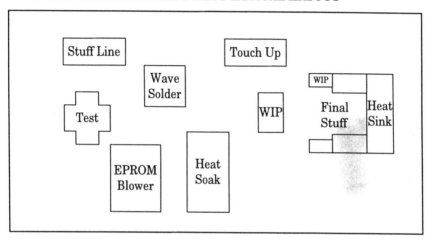

It is useful to now superimpose the actual *material* flow on the layout diagram to get a view of the extent of the problems associated with this layout.

FIGURE 2.6: ELECTRONICS FIRM MATERIAL FLOW

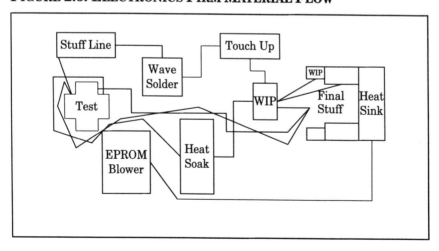

Product crosses back and forth over itself throughout the factory. The level of supervision required to monitor, control and progress such an operation is quite high. To address this situation it becomes important to go back to basics.

"What Are We Trying to Do?"

The layout shown in Figure 2.7 was suggested as a first step in improving the physical flow of the operation and so improving the productivity and responsiveness of the system.

FIGURE 2.7: RECOMMENDED PROCESS FLOW

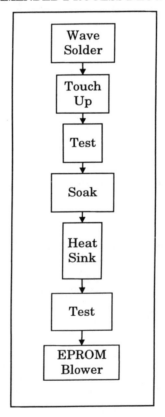

The flow of product has been improved substantially. There is significantly less space for work-in-progress to build up. The physical layout itself helps to control the process. Management of this physical process flow is also simpler as parts follow the logical process flow sequence. This concept of physical flow will be further developed in the following section on cellular manufacturing.

A Case Study

Whereas the previous example related to the electronics industry, this next example relates to a more traditional engineering industry.

In engineering industries the traditional layout arrangement has been along process-oriented lines, i.e. work is carried out in particular departments or areas. All the sawing would be done in one area, the milling in another and the turning in another under the supervision of departmental managers. Once again, to determine areas of waste in the company in the case study, an analysis of the manufacturing process is needed.

The original layout in this example is presented in Figure 2.8 The diagram also contains an outline of the material flows and thread diagram for the plant (this is not the full material flow diagram, only a summary of it). No details of work-in-progress at each machine is given.

FIGURE 2.8: INITIAL LAYOUT

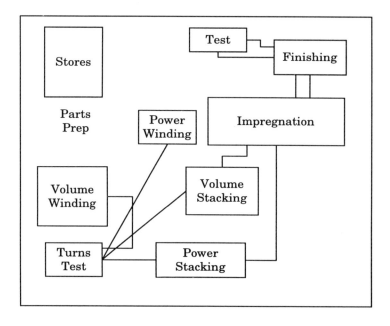

The company manufactures two distinct ranges of products. These were simply differentiated as being Large and Small with different handling and processing requirements for each product group.

An initial analysis of the operation indicated the "necessity" to build an extension for the factory. Before the company embarked on this expensive course of action they decided to try the WCM way to see if they could avoid this costly investment.

Two lists were made of the *actual* steps involved in the manufacture of the products. These lists consisted of full details of the operations as outlined below.

Production Steps for Small Transformers (existing):

1. Route card issue
2. Requisition or get materials
3. Winding
4. Turns test
5. Bank and complete windings
6. Turns test
7. Tray to WIP
8. Coil finish
9. Supervisor decides and organises
10. Solder
11. Cosmir set-up
12. Get/requisition materials
13. Stack, into tray
14. From tray, fill in and close
15. Supervisor, organise, decide
16. Dry test set-up, from tray, test, to tray
17. Move to WIP area
18. Varnisher moves from WIP to varnish, from tray, to varnish to oven (2/time)
19. Next AM, supervisor advises clamper, clamping
20. From oven onto trolley, to clamp into tray, fit clamp and punch, into tray
21. Move to finish area
22. Get materials from Eubanks
23. Remove from tray, solder, put into tray
24. Supervisor decides crew for sleeving and shrouds and moves staff
25. Sleeve and shroud
26. Wait
27. Move to final test
28. Supervisor decides when to be tested
29. Final tester finds route card
30. Set-up for final test
31. Remove from tray, test, place in tray
32. Documentation
33. Move to shipping area
34. Packing, get materials
35. Pack

Production Steps for Large Transformers (existing):

1. Find route card from Tray, Supervisor, Prod. Mgr, Engineer
2. Think and evaluate priority on job
3. Pick materials or wait for delivery
4. Set bobbin up
5. Cut nomex
6. Wind primary
7. Make screen and fit
8. Wind secondary
9. Move to turns test, test and return
10. Coil finishing, fold/unfold solid wire
11. Nomex
12. Decide flow
13. Move to power stack
14. Supervisor decides action
15. Fold leads out
16. Get materials, wait, drill
17. Hold workpiece (two workers)
18. Finish stack and assembly
19. Fold wire and lift to trolley (two workers)
20. Dry test
21. Strip leads
22. Set up meters
23. Route card search
24. Test
25. Lift to trolley
26. Move to varnish area
27. Lift to varnish tank (two workers)
28. Lift out of tank and into oven (two to three workers)
29. Bend leads
30. Carry to finish area
31. Decide next job
32. Change decision
33. Get materials
34. Search for drills and spanners
35. Pull and cut leads
36. Finish
37. Lift to table
38. Wait for final test
39. Next for final test
40. Final test
41. Move to table and wait
42. Decide packing method
43. Get packing materials
44. Get pallet
45. Pack

Figure 2.9 shows the theoretical steps involved in the manufacture of the products.

The difference between the production steps for small and large transformers is obviously the starting point for the development process.

FIGURE 2.9: THEORETICAL STEPS IN TRANSFORMER MANUFACTURE

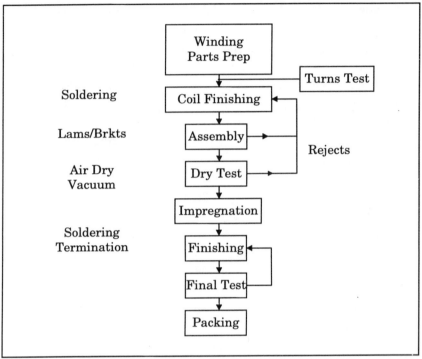

The wastes in the existing production process were identified and steps taken to remove them. Product groupings were identified and made universally known. Small transformers were known as "volume", large transformers were known as "power". Machines which were required for specific product groups were brought together and put closer to each other. Supervision of groups of people was delegated to group leaders. Material flows were reduced significantly by the simple expedient of looking at what the company was trying to manufacture rather than satisfying a departmental structure. The interim layout shown in Figure 2.10 illustrates how the finishing of the products became more focused. The large transformer finishing was grouped into a distinct flow line; while for the small transformers the company found that by grouping three operators together they could improve productivity by 50 per cent over the previous situation where they worked independently, creating as much WIP as finished products.

FIGURE 2.10: INTERIM LAYOUT

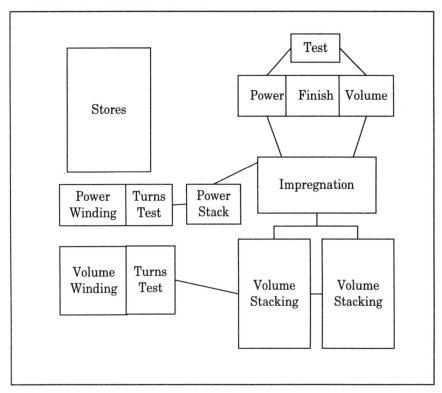

This was an interim stage in the development process. Local management and staff had to come to understand the potential benefits and difficulties they would encounter when moving towards cellular manufacture and self-directed teams. As minor local successes were achieved both management and staff became more comfortable with the idea of World Class Manufacturing and how it related to them.

A key point to note is that these developments of the process were made as the company continued to service its customer base and actually during a period of business development. The development process continued and the present layout of the plant is presented in Figure 2.11.

FIGURE 2.11: CURRENT LAYOUT

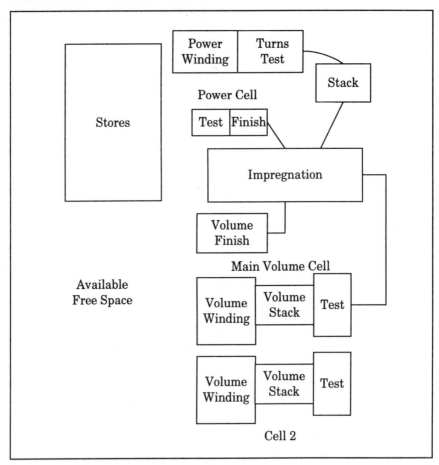

The process of cellular manufacture is well developed. Products have been grouped and the machinery arranged to facilitate their manufacture. When one worker has completed their task they pass the part directly to the next worker's in-tray. This has effectively removed the vast bulk of the work-in-progress from the process. People working in cells have taken responsibility for their own output and will proudly display their achievements in terms of quality and quantity of production. The flow lines for materials approximate very closely with the theoretical steps. The production is effectively self-managing; as soon as a job is started at the beginning of a cell it will be completed within a 24-hour period. The management of the process is simple and highly visible.

The main advance made in these examples was from asking the simple question: "What are we trying to do?" rather than looking for sub-optimisation at departmental levels.

Why Change?

The traditional departmental system seems to make sense. A supervisor, charge hand or manager was responsible for each specific area. That person could control and report on that one area and the process was under one individual's control in each area. When we look at the total operation we begin to see the problems. Wastes begin to become visible:

- *Excess WIP* to "allow" machines and operators to be "busy" all the time.

- *Difficulty in managing the overall operation.* Where is a given order at a given time ?

- *Long lead times.* Each department has its own lead time, to work through its backlog of WIP, before it can process any other order.

- *Double, treble and quadruple handling* of parts, intermediate storage, stock handlers and checkers.

- *Difficulty in tracking down the true causes of defects* — for example, with parts in WIP which were generally produced 1 to 5 days before being used. Was there a problem with a particular machine, material or person when they were produced? How can this be ascertained days after the job was completed and forgotten about?

These are just some of the wastes that can occur in a departmentally organised operation. Frequently there is a distinct lack of ownership and responsibility between different departments and communications can often be tenuous and difficult. Supervisors, operators and managers tend to work to maximise their own efficiency as measured by parts produced rather than as a holistic approach on finished products at the end of the line.

The examples outlined above show steps along the way to cellular manufacture.

CELLULAR MANUFACTURING

The previous example of the electronics manufacturer brought the physical and process flow concepts out for discussion. An essential point to be understood when dealing with WCM implementation, however, is that you are never finished critically analysing your operations. We will return to the electronics example to demonstrate what this means.

The earlier example showed a clear improvement in the areas of WIP, supervisor input and control time as well as physical space requirements. One key point not to be missed was that this layout only addressed *part* of the total process — it did not look at the initial stages of the part stuffing, i.e. placing of the electronic components onto the boards themselves.

The company decided to move to look at the total process, using the cellular manufacturing approach shown earlier in Figure 2.4. It was manufacturing products in batches of 200 to 500 units. The stuff line had to be set up for each product, with kits being pulled from the stores. Frequently, sufficient stocks of parts were unavailable to allow for completion of batch orders. Problems arose in trying to manage the stores area. Quality was measured in terms of first time pass rate (FTPR), i.e. when all the parts were assembled and the electricity turned on, did it work? Industry quality figures range in the 70 per cent to 80 per cent FTPR.

The cellular manufacturing approach starts by identifying "product families". These are not necessarily groups from the same product range, but rather groups of products with similar parts, size and processing requirements. Analyses are carried out on the parts requirements for the family. Availability of marketing/selling forecasts are used to determine initial capacity requirements for the cell or cells.

One of the first problems encountered when trying to create a cell is determining the number of people required to work in it. This can be determined once the sales requirements have been evaluated or at least estimated. We also need to have a good idea of the amount of time required to manufacture the product. We can determine the number of staff required to work in a cell by using the following formula:

$$\text{No. of Cell Staff} = \frac{(\text{Manufacturing Time} \div \text{Part}) \times (\text{Qty Required})}{(\text{Working Time Available per Worker})}$$

In Figure 2.12, the wave solder machine has been located at the centre of the layout. Products have been grouped into cells with the flow of materials moving from the stuff area, out to and back from the wave solder machine, through test and on to final assembly and pack. The work required to manufacture a product is concentrated in a single area.

FIGURE 2.12: WCM CELLULAR MANUFACTURING LAYOUT

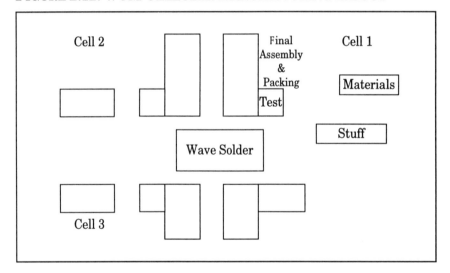

The main benefits accruing from the arrangement are:

- Quality improved by 15 per cent. FTPR is now over 95 per cent. The close proximity of the workers has meant that problems are handled almost immediately. The cell members go to tea together, there is no desire to *not respond* to a colleague's legitimate concern.

- WIP is reduced by 75 per cent. There is very limited space to hold WIP. There is also no need to do so as work is performed as it presents itself.

- Lead time reduced from 4 weeks to 1 day. Because the complexity of the operation has been removed, lead times have been slashed. Lead time has approached actual processing time. Wastes have been removed from the production process.

By focusing on the specific job, waste can be eliminated. That may sound simplistic, but simplicity is the key to a successful WCM implementation. Search for waste, remove it, make the job easier and simpler and improve performance and throughput.

Lead Time Reduction

One of the benefits of a cellular manufacturing approach is that lead times can be greatly reduced. This is most easily understood by looking at an example. Let's assume that in order to manufacture a certain product a square of material has to be drilled, shaped and sawn. The customer wants three pieces and each of the actions — sawing, shaping and drilling — will take one hour to complete per part (we won't try to measure transport times or any waste times in the process).

If we were to process the order in a traditional way, the parts would pass through each operation in a batch of three, just as the customer ordered. In that case the elapsed time to complete the order would be 9 hours. And if we were to look at the amount of work-in-progress (WIP) waiting before each machine we would see that we can have up to 3 pieces in WIP.

Now, let's assume we were to use a World Class Cellular Manufacturing process. We would still want to produce the three items the customer ordered, but the parts would pass through the process one at a time. As soon as one operation is finished on a part, the next operation could start on it. In this case the elapsed time would be 5 hours, with only two pieces of work-in-progress in the system at any time.

We can see how the overall lead time can be reduced by adopting a World Class Cellular Manufacturing system, but there are other advantages of the system demonstrated in these examples. In the batch processing system it would be seven hours before we had the first saleable item from the process; the World Class system has saleable product after three hours. If we use a number of other World Class techniques in the process, such as failsafing systems, we can potentially reduce quality faults in the system as well, as we will see below.

The first operation in the process is to drill the plates. If the second operation uses this drill hole to help locate the workpiece then an automatic check is made to see:

1. If the hole is present.

2. If it is the right size.

3. If it is located in the right place.

A situation can arise where the drilling machine is set up incorrectly with the wrong drill bit, or in the wrong place, or the drill can break during use. The batch processing system will potentially allow at least three items to be manufactured incorrectly before the problem is seen. It would be more typical for the full three steps to be completed before such an error was detected. The World Class system only allows a single part to be produced incorrectly before the problem is identified and corrected. The question of lead time is purely arithmetic. If the batch size was increased from 3 to 5 items, for example, then the processing times for the two production systems would be:

	Elapsed Time	*WIP Levels*
Batch Processing System	15 hours	5 units
World Class Cellular System	7 hours	2 units

We can work out the expected processing times and WIP levels for larger batch sizes quite easily. In a World Class Company, lead time and WIP are seen as serious wastes. They cost the company money and prevent it from responding quickly and effectively to customer needs. Every effort needs to be made to reduce them to the minimum.

By reducing the batch size and passing items on as they are produced, by focusing on the finished product rather than sub-components or sub-assemblies, we can produce products quicker with less capital tied up. The focus is on overall optimisation rather than departmental sub-optimisation.

OTHER INDUSTRIES

The electronics and automotive industries have been the front runners in implementing JIT and WCM across the world. The first comment that people make when they hear about WCM, however, is usually: "But that wouldn't apply in our industry . . ." An essential point to understand is that the basic principles of WCM are not restricted to these industries — they are universally applicable.

FORECASTING

A forecast of sales allows a company to plan for its production, its materials acquisition requirements and its financial arrangements. It is logically a key and central element of a business, or at least it should be. Many companies prepare a budget or business plan for an external body such as a bank or state agency, yet few SMEs maintain such forecasts and fewer still monitor their accuracy or work at improving them over time. The effects of "poor forecasting" are readily seen:

- No stock to fill the many back orders, despite the fact that the company has large stocks of finished goods on hand — unfortunately they are the wrong ones

- Purchasing unable to obtain parts from key suppliers due to long lead times

- High inventories of raw materials, work-in-progress and finished goods

- Factory working large amounts of overtime to rush through jobs

- Half finished products around due to parts shortages.

Let's look at a real-life scenario.

Company A is suffering. The Sales Department was unable to fulfil orders last season because it could not get sufficient products from the Manufacturing Department, who couldn't make the products because the Stores Department hadn't got the parts in stock because the Purchasing Department hadn't ordered them. Why, one may ask, had the Purchasing Department not ordered the parts — they had a forecast, didn't they? Well, in fact, they didn't, because the Sales Department didn't know what their customers would want to buy so the Purchasing Department had to make a best guess. In this case, there were a number of further complications. The product range was very large, with approximately 50 models with up to 20 plus options on each. The suppliers for some of the major components had real lead times of three months.

The solution to this case (which does exist and, though it appears to be extreme, is actually fairly typical) started with a shift in primary responsibility for deciding what the customer would buy from the Purchasing Department to the Sales Department. In actual fact, the full management team of Sales/Marketing, Production, Stores, Purchasing, Finance and Research & Development came together to analyse and rationalise the product range and the options available. This led to a reduction from 50 models with 20 options to 12 models with five options. The problems for the team had been immediately reduced. With this new product range, the sales staff approached their customers to find out what estimated quantities they would require for the next season. They ended up taking firm orders for the first two months of the new season, with estimates for the third. These forecasts then allowed the purchasing and production staff to prepare for the new season based on a relatively good idea of what was likely to be required.

A key point to be noted in relation to forecasting is that the process needs to be ongoing. The people with primary contact with customers should be the ones generating the forecasts. Feedback needs to be given to monitor the accuracy of the forecast, and this feedback will in turn have the effect of helping the forecaster to be more accurate.

Typically, approximately 20 per cent of orders forecast for a production unit are accurate to plus or minus 20 per cent. Forty per cent of orders are not forecasted and 40 per cent of orders forecasted are not produced or produced at less than 80 per cent of the forecasted level. The impact on a business of this level of forecasting accuracy can be significant:

- Raw materials become unavailable

- Stock holdings escalate

- Overtime working levels increase considerably.

These are all seen as operational problems, and they are ultimately handled by the operational people, but the root cause of the problems rests with the full management team. The primary impact of poor forecasting is on the working capital requirements of the company. The problem needs to be addressed in a number of ways simultaneously:

- Identify the problem for what it is

- Improve contacts with customers to reduce inaccuracies in forecasts

- Reduce internal lead times and improve speed of response of production systems

- Work with suppliers to enable improvements in lead time along the supply chain.

By bringing these elements together in a focused way the basic working capital requirements of the business can be reduced. Without accurate forecasting it is very difficult for purchasing and manufacturing staff to supply products to customers when they want them.

PRODUCTION CONTROL SYSTEMS

The traditional approach to production control is to take an order, place it in a planned queue and manufacture it when it comes to the top of the queue. This seems perfectly reasonable, and it is. We have seen already the use of cellular manufacture and how this effectively allows for a multiplicity of queues as it puts in place a number of dedicated production units focused on a single product or product family.

The World Class company looks at production control in a different way. The importance of orders are taken into account when planning production and the type and style of order is relevant. If a product is a mainstream, long-running product with significant production volume then it should be manufactured and controlled in a different way to a product that is a once-off or low volume item. The World Class company uses the Pareto Analysis tool to help identify which categories a product fits into. Product volumes are typically analysed to identify which products or variants are the high, medium and low volume sellers.

A typical breakdown in a manufacturing plant shows that a small number of products account for a large proportion of the volume. Many companies manage the order for the small volume products in the same way as for a high volume product. The World Class company, on the other hand, tends to use three basic systems of production control, depending on the volumes to be manufactured:

- Make to Order (MTO)

- KANBAN

- Rate-based Scheduling.

Make to Order is suitable for low volume, highly unique products. KANBAN is best suited to middle volume, middle difficulty products, while Rate-based Scheduling is suitable for products to be produced in high volumes with little diversity.

Make to Order

As the name implies, Make to Order (MTO) means that products are made as an order is received. Basic stocks of parts are held at a minimum, if at all, and products are basically "specials" and handled as such. Management control these jobs as individual projects and they can often be demanding of time and resources. A company should be aware of this and know exactly why they are producing these products. By isolating the costs associated with their manufacture, management may often query why they are offering them for sale and at what price. Often these products must be offered to complete a product range or satisfy a specific need of a good customer or for some other strategic reason.

By using the MTO production system these specials can be separated from the normal run of production, reducing the disruptive effect they have on normal production items. Also, as the inputs and costs associated with their manufacture have been identified and highlighted, management has the opportunity of considering whether or not the manufacture of these special, low volume items is worthwhile for the company. Often it may be more suitable to have these parts or products manufactured on contract by an operator who deals specifically with specials and low volume production.

KANBAN

The KANBAN system is typically a Two Bin system. Products are manufactured up to a given quantity, and when sales have drawn down from stocks to a certain level, a re-order request is made on production. The system uses simple re-order triggers such as a card with standardised production quantities and product details on it to simplify the replenishment process. Materials are generally kept on hand at reasonable levels and suppliers are kept updated as to usage rates on an ongoing basis. The system is effectively self-regulating, with the cards or bins cycling more quickly or slowly, depending on market demands. Annual or monthly forecasts are used by purchasing to keep control of ordering requirements and to support suppliers.

Rate-based Scheduling

The Rate-based Scheduling system is applicable for high volume products that have a definite and secure place in the market. The production system is set up to manufacture at a steady rate and production volumes are forecast and "plugged into" the manufacturing system. Materials are ordered according to these forecasts and suppliers know what they need to deliver on a regular basis. The system is particularly applicable where a steady market state for a product has been achieved. The production unit and its ancillary support services effectively go on fully automatic operation. As product volumes increase the production process becomes an ever more likely target for automation.

Figure 2.13 illustrates the applicability of each of the three systems.

FIGURE 2.13: PRODUCTION CONTROL SYSTEMS

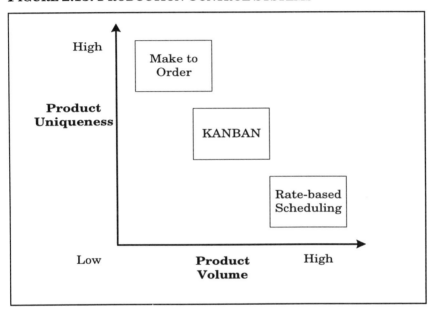

Product Life Cycles are another important factor for consideration when choosing the appropriate manufacturing control system for a given product. The Pareto-ABC analysis discussed earlier is accurate at a given time. The relative importance of a product and its related sales volumes is likely to change with time. As products move through their life cycles they will logically move from one system of production management to the next, as shown in Figure 2.14.

FIGURE 2.14: PRODUCT LIFE CYCLE

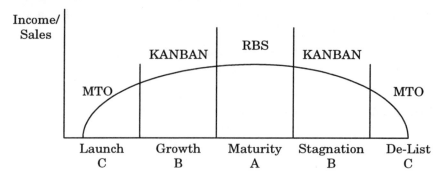

Management need to be aware that a given production control system is most suitable for a certain time in a product's life cycle and should change the production control system as appropriate. Imagine making mint sweets one at a time! Once a product has been placed in a given production control system, the system itself looks after the day-to-day control of its production, controlling the ordering and manufacturing sequences.

SET-UP TIME REDUCTION

To allow us to compete effectively we need to be able to change-over and set up our machines with the optimum efficiency. A short case study may best illustrate that fact.

A company located in Dublin, looking for growth possibilities, targets the baby consumable sector. Marketing studies identify an opportunity for a heavily branded premium product to compete with the market leader, a major multinational. Technology is decided upon as the means of obtaining differentiation in the market, and a suitable machine is located in the "home" of advanced technology — Japan. Then the problems start. . . .

Japanese firms tend to consider machinery differently than do those in the West. Basically, they have a preference for "good" machines rather than "new" machines, preferring machines with simplicity and proven capability. In the West we too often think we need all-singing, all-dancing super machines to solve our problems. In this case, for example, the machine the Dublin company wanted would have to run three shifts and be capable of producing product from the smallest to the largest size. In Japan, on the other hand, companies usually run their machines for two shifts and spend the third shift maintaining them, doing development work or changing over from one size to another. The problem is immediately obvious.

Very few production lines or machines can be set up for a specific product and left that way. "Changing over" is a generally commonplace occurrence in industry. Once again, the Japanese took the lead in this area. Where Western industry has moved towards super machines that need to be run at maximum speed practically all the time to cover the costs of investing in them, the Japanese have moved towards simpler machines. These generally have lower levels of automation that are in themselves relatively cheap to finance and consequently can be run as demand requires. This flexibility allows for the shift from a process-oriented to a product-oriented manufacturing operation with all the previously identified cost savings.

Change-Overs

Production managers, supervisors and machine operators have a distinct dislike for change-overs. The lost production time and the lost product as the process is re-started are both headaches. A production manager usually would like to produce enough product at a given time to last for months. If we follow this logic we end up with large inventories of WIP with all the problems of storage, handling and quality associated with it.

As noted, however, the marketplace is changing. Demands are increasing on sales staff to be able to offer products to customers quicker and often with a wider variety. This change affects production staff dramatically.

Reducing set-up time needs to be approached in the same way as any other WCM problem. An analysis needs to be carried out of what actually happens during a change-over. This analysis usually consists of a list of actions which forms the basis for further development work. The use of video recording equipment can often facilitate this analysis phase by ensuring that seemingly insignificant actions are not overlooked in the overall analysis.

The standard classification of change-over actions is whether they are *internal*, that is, they can only be performed when the machine is stopped, such as replacement of tooling, forming sections, moulds etc., or they are *external,* that is, work that can be carried out while the machine is running, such as getting the parts and tools ready, initial settings, materials location and preparation, general supporting work, etc.

The use of brainstorming activities and team problem-solving is very useful when it comes to set-up time reduction. It is a classical point of cross-over between the traditional industrial engineering approach and the "new age" JIT techniques. The use of standardised tooling, quick release mechanical and electrical connectors, standardised datum points and a host of other detailing can have a significant impact on overall set-up time. Common sense is also a very useful tool in this area.

In an Irish-based food multinational, the use of set-up time reductions was very helpful. Simple little things can have a major impact on the process. The first stage of the process called for organisation of the workplace. This entailed the use of simple racking, using stillages, to allow the parts and components for each of the major machine changes to be grouped together logically. The helper for the line could have the pallet boxes of parts out of the racks and ready for the fitter as the line staff were washing down the production line.

All electrical connectors were developed to be simple quick disconnect connectors. All special tooling was kept with the specific parts they related to. Hand release locking devices replaced ones requiring tools. Parts removed from the machines were cleaned and checked and repaired if necessary, before being put back into the storage area.

These simple, common-sense actions greatly improved change-over time by removing complexity from the operation, by having parts ready and available when they were needed, where they were needed. The basics were taken care of.

This simplicity of approach needs to start at the design stage of the product. The most effective set-up time reduction is to avoid the necessity to perform a change-over in the first place.

The steps to reduce set up time can be summarised as follows:

1. Analyse what is internal and external.

2. Shift work from being internal to external.

3. Reduce and simplify internal work. Improve attachments, reduce settings etc.

4. Reduce overall set-up time.

When a standard method of performing a change-over has been identified, the use of the video recorder is again useful as a training aid. The new procedure can be recorded and used to train all staff members to the new method.

The range of potential improvement tricks, methods and techniques to improve set-up time is very large and is generally accepted to be very well documented. The following examples are presented as indicators or stimuli to the type of work carried out during a Set-up Time Reduction Programme:

- Standardisation — of dies, baseplates, connectors, fittings etc.

- Clamping devices — to remove the necessity for bolt tightening

- Trolleys and carts — for easy transportation of heavy parts

- Modified bolting arrangements — where slots, grooves, modified washers etc. are used to retain parts

- Locating devices/pins — to locate parts accurately and easily.

STOCK ANALYSIS AND MANAGEMENT

Very few manufacturing operations have only one part in their products. Most operations have from 100 or so stock items up to many thousands. Even the smallest of craft-based companies can have a wide range of parts and pieces to monitor and control.

Traditional stores control systems have either paper or computer-based systems where stock inputs and outputs are recorded. Materials Requirement Planning (MRP) systems are quite widespread. For such systems to work, however, a stock control accuracy of over 90 per cent is required. This demands a high level of control and basic work — none of which adds value to the parts.

In a WCM environment the total focus is on simplicity and adding value to products. Any actions or work input that does not add value is waste, a cost addition to the finished product and thus something to be avoided. The WCM factory endeavours to remove as much of the cost of stock control as possible. The biggest and most compelling reason for a stock management system is to ensure that parts and components are available to manufacturing as and when required. Economic considerations preclude holding very large stocks. But in a WCM environment the lead time has been reduced significantly, large amounts of WIP have disappeared and effectively parts are used when they arrive at the factory. In one of the Japanese automotive plants visited recently, suppliers' delivery vans were seen delivering directly to the line! JIT and WCM in most Western companies still has some way to go before these practices are widespread.

If we start with the primary function of stores as being to ensure that parts are always available to production, we begin to get a handle on the problem from a WCM point of view.

Production schedules are based on sales forecasts, which in turn define expected stock movements. When sales forecasts are analysed using the Bills of Materials a view forward can be taken as to the future usage of parts by production. This tells us what we estimate we will need to purchase for a given time period. It now becomes important to examine the financial part of the stores operation. The high cost of holding large stocks of parts was mentioned earlier and the use of ABC analysis to manage this cost is now examined.

ABC Analysis

The use of ABC analysis can be a useful tool in controlling parts. An ABC analysis consists of examining the usage of parts in relation to their stock valuation. Typically, each stock item is assigned a cost which is the annual quantity used times the stock item price. The stock items are then sorted by annual valuation.

For example, Class A items are defined as those whose annual valuation totals 80 per cent of the total purchases, Class B items account for about 15 per cent of purchases with Class C items accounting for the remaining 5 per cent. This analysis usually results in a definition of how parts should be handled and controlled, with a high level of monitoring taking place on the Class A items with reduced attention being given to the B and C Class items. This generally results in managed stocks from a financial viewpoint. However, remember that at the outset we said the *primary* role for stores was to ensure that parts were available to production as required.

The WCM view focuses on production and ensuring that production can be sustained, but with an economic awareness as well since waste of space and money in holding stocks is frowned upon by the basic WCM principles. Thus the WCM ABC analysis looks at the financial/annual sales generated analysis in terms of delivery time, i.e. the time taken to replenish stocks.

The annualisation of sales forecasts with the Bill of Materials information is used to determine annual requirements against which general purchase orders are placed. This system allows suppliers to plan *their* production to ensure availability of parts to the purchaser. Parts are not "called in" until required. This shift in stock management can bring a number of difficulties to the material control department.

1. They need to know, accurately, when to call in parts.

2. As the frequency of call-ins increases, so the system of calling in parts needs to be simplified.

3. Simplification and streamlining of the total materials and stock control systems are required.

It is critically important to develop the material management system to ensure that parts are available to production as required.

Two Bin System

A lot of time is spent counting parts in stores to check how many are there or how many are to be issued against a specific parts requisition. Once again, these counting operations add *no value* to the product, only cost. In a World Class Manufacturing environment, this counting can be reduced significantly.

The two bin system is an extremely simple system of stock control. Parts are stored in either of two bins. Parts are used from the first bin and when it is empty the bin itself acts as a stimulus for replenishment. The bin can be a physical bin, a stores location, a bag of parts or even barrels or vats of material. The use of the term bin is only for convenience. The following example demonstrates the idea.

Bin A contains a chosen number of a particular part. The bin quantity is generally decided upon as being sufficient to meet a number of days' production requirement initially. As the system is proven, the bin quantity can be reduced to whatever is sufficient for a number of hours' production. When bin A is emptied it is left in a collection area for the materials staff to fill it. The bin usually carries a label identifying the part, part number, quantity etc. on it. The manufacturing staff proceed to use the parts in bin B while bin A is returned to the stores.

At the stores, bin A is replenished from a larger bin, C, and is then returned to the manufacturing area, for the cycle to be repeated. Sizing of the contents of bins A and B is such as to strike a balance between frequency of replenishment and physical space requirements in the manufacturing area.

The system in relation to bins C and D is very similar, where the quantity held in each bin is sufficient to allow for replenishment of the bin from the supplier. As the system develops, bin D can physically exist in the supplier's premises thereby reducing stocking requirements for materials.

A particularly interesting feature of the system is its simplicity. Control can be carried out on a purely visual basis. There is no counting to see when bin A will run down. There is no matching of production forecasts on a daily basis to check if there are sufficient parts in the cell to manufacture the day's requirements. If the bin is empty it needs to be replaced and the full details of what and how many it contains are clearly visible on the bin itself.

Many materials specialists initially have difficulty with the system, but it works and they generally come to terms with this simple fact. It allows materials people a window of opportunity to obtain stocks before they run out, rather than having to manage many hundreds or thousands of parts on a daily or weekly basis. This can mean a significant reduction in workload and time for staff — time which can often be better spent developing relationships with suppliers, in looking for and identifying alternative, possibly local suppliers who may well be able to supply requirements *as needed*, thereby minimising stock holdings still further.

In the early stages of implementing a two bin system, stock holding valuations may rise. As confidence builds in the effectiveness of the system the bin quantities can be reduced as supplier performance improves or alternative better quality sources are located. Supplier development is a key element in a full implementation of WCM within a company. As suppliers themselves take on the concepts of WCM their lead times and quality will improve to the shared benefit of all (see Chapter 8, "Supplier Development"). As the two bin system is being introduced, however, it is better to err on the positive side to avoid stock shortages *due* to the new system. Such shortages would obviously have a bad impact on how the system is perceived.

TOTAL PRODUCTIVE MAINTENANCE

Western engineers like to be busy. Accountants like machines to be busy. Japanese companies, on the other hand, plan their day to be busy 60 per cent of the time.

A previous example related to the purchase of a Super Machine. This machine was required to run 24 hours a day, 5 days a week to cover the cost of its purchase and to meet the marketing department's forecasts. The machine had been designed, however, to run for 16 hours a day with 8 hours maintenance, fettling, cleaning and adjustment. The problems caused by these different expectations need to be examined in some detail.

Japanese companies believe in good machines, ones that have been sorted out and made to work well. Engineers have found that machines perform according to the bath tub principle, so-called because of the shape of the graph shown in Figure 2.15.

FIGURE 2.15: PROBABILITY OF FAILURE VS. TIME

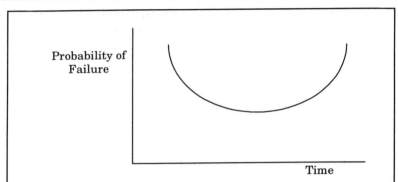

When a machine is new, it tends to break down until any teething problems have been sorted out. It will usually then have a useful life after which problems will start to re-occur. By applying Tender Loving Care (TLC) or planned maintenance and development to the machine, the length of its useful life can be extended, theoretically, almost indefinitely.

In the small company with the super machine, there was also a good machine, valued at less than one-fifteenth the price of the super machine. It ran at 100 pieces per minute and performed at the rate of 95,000 to 110,000 items produced per shift, regularly, for three shifts. The super machine rarely managed to produce in the 80,000+ range.

The good machine was understood by the skilled team of fitters and operators. The level of technology was appropriate to the job. The equipment was hardened, toned after a number of years of on-site tweaking and development. Chains and shafts drove the separate elements of the machine.

The super machine, on the other hand, was highly advanced but over-complicated (this author spent three days in Japan with design engineers trying to find a solution to a particular problem involving six double-sided electronically-controlled power clutches, to no avail! On his return to Ireland he disconnected three of the clutches completely — the problem was solved). The components of the machine were sequenced by rubber timing belts and a myriad of special gearboxes. It ran at speeds of up to 350 units per minute — but not for too many minutes.

The point of the above example is this:

- Use the appropriate level of technology for the problem in hand.

- Understand the technology before you buy it!

- Know what maintenance will be required.

Over-producing or running a machine too fast can often be a costly exercise. This is the first point to be understood in relation to Total Productive Maintenance.

Maintenance Management

The basic requirement in a WCM factory is not that machines are always available to work, but that they are always available to work when they are wanted and that they will work at the output required of them. This does not necessarily mean at their design speed, as shown in the example above.

Traditional maintenance management is based on the following five categories of maintenance:

1. Fixed-time maintenance

2. Condition-based maintenance

3. Operation to failure

4. Opportunity maintenance

5. Design-out maintenance.

The particular maintenance type chosen for a given plant depends on the type of plant and the type of equipment involved in the process. The maintenance plan for a sewing factory where machines can be replaced very quickly and easily is very different to that employed by an electric utility where an electricity generation plant is concerned and machines can weigh over 150 tons. The following section will discuss these different philosophies of maintenance.

Fixed-Time Maintenance

The fixed-time maintenance method is well known to anybody who drives a car — you replace your oil, spark plugs and air filters at given time periods. The manufacturers have determined the time after which the items are suspect so you change them before they are likely to cause a problem. Similarly, certain checks on other key components, such as brakes, are made against a defined time schedule. The logic behind this system is that parts and machines have predetermined service lives with a high probability of parts performing faultlessly during this period. After the life cycle, the probability of a failure increases. The fixed-time maintenance idea is based on sound experience and is represented in Figure 2.16.

FIGURE 2.16: COST OF REPLACEMENT VS. FREQUENCY OF REPLACEMENT

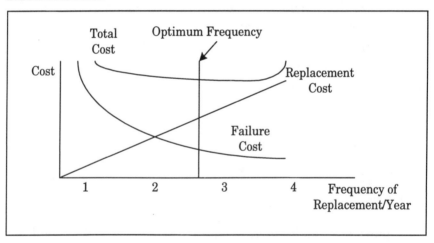

Figure 2.16 represents the cost-based logic behind many parts replacement decisions. There is obviously a cost involved in replacing parts, the cost of the parts themselves and the cost of doing the work. There is also a cost associated with allowing the parts to fail in service, presented as the Failure Cost. The more frequently we replace the parts the lower the Failure Cost. The Total Cost Line represents the summation of the two costs. The Optimum Frequency line is located at the minimum Total Cost point; this would be the optimum frequency of replacement point

to minimise total costs, in this example 2.5 times to change the parts per year.

The major difficulty with this system of maintenance is obvious: it is often a complex problem to get enough statistical data to determine the failure probability characteristics for single parts let alone for complex machines. Complex machines can have a large number of individual parts, each with their own failure mechanisms. This makes a time-based maintenance system unsuitable for such machines. However, oils, greases and filters are simple items which have easily determined operational lives, making the system appropriate for them.

Condition-Based Maintenance

Condition-based maintenance is probably the simplest, most effective and most widespread type of maintenance used in manufacturing. In its simplest form it consists of the engineer, fitter or operator using their five senses to care for a machine. A number of these actions can seem very simplistic:

- Wiping a surface

- Listening to a machine

- Smelling

- Looking at a mechanism's action or a shaft's rotation

- Feeling for vibrations

- Tasting (yes, tasting) fluids or products.

An experienced operator or engineer can often predict failure of a machine long before it leads to a loss of production. Operators, particularly, are the cornerstone of this maintenance system. As they spend so long each day using their machines they quickly come to know when something is going wrong. Personal experience working with sweet packaging machine operators in Nestle taught this author the great benefit of listening to machine operators. They often can't explain technically what is going wrong but they are usually more aware of changes in machine performance. This interaction between technical support staff, engineers and machine operators is an essential element in Total Productive Maintenance.

As machines and processes become ever more complex, diagnostic tools and systems have developed to allow for data recording and analysis. Such tools can be very useful but are often complex and expensive in themselves and usually fall within the category of "engineers' toys". The old 80/20 Pareto rule still applies: even though a machine may be very complex, its basic functions are usually quite simple. Apply your senses and experience to solving problems and you will generally solve it. The additional diagnostic tools may help you confirm your diagnosis.

Opportunity Maintenance

Opportunity maintenance relates to the parts and items you would fix when something else breaks down or is being dealt with under a planned or condition-based maintenance programme. Generally there are parts that may wear at a reduced rate to the failed part that can be replaced at maintenance time, thereby giving an overall reduction in maintenance costs as the machine will not have to be stopped and stripped down to replace parts on two separate occasions. The decision as to what additional work gets done is generally based on experience. An example is replacing springs in a sealing machine when belts are replaced, or doing an air filter replacement when the spark plugs are being replaced.

Operation to Failure

This maintenance plan does what it says: it runs an item of plant until it breaks down. At that point it is either repaired on site or replaced completely. This type of maintenance system has only a limited application in a WCM factory. If the operation to failure system is operated in conjunction with a condition-based monitoring system the plant can be run to failure. Management can plan for rapid replacement of the failed machine or part, thereby ensuring that machine availability is maintained. The balance between condition-based maintenance and operation to failure is a fine one with the economics of the situation being the deciding factor.

Design-Out Maintenance

The last of the traditional maintenance systems is that of designing out maintenance. This often takes place in the factory itself. Machine designers traditionally design their machines at a distance from the factories where they are used. Local engineering staff tend to improve and develop machines based on their operational experience. Quite often these developmental improvements can have a significant impact on the performance and productivity of the machine. Once again, economics decides whether or not a re-design takes place or the continuing costs of ongoing maintenance are suffered.

TOTAL PRODUCTIVE MAINTENANCE IN AN SME CONTEXT

As discussed previously, companies in the West and the Far East differ significantly in their attitudes towards machines and maintenance. In the West we have traditionally focused on super machines, running for 24 hours a day, with a jaundiced attitude towards maintenance and the maintenance department. The maintenance department is usually seen as a necessary evil. A cost. Senior management have, by their actions, shown scant regard for the importance of proper machine care and development.

In the Far East, on the other hand, maintenance of a machine is viewed in a very positive manner. Machines tend to run for 16 of the 24 hours with the non-running time devoted to cleaning, oiling, adjusting and maintenance. A simple walk through the Nissan Diesel Plant in Japan contrasts quite starkly with a similar walk through the Scania plant in Sweden. The Japanese machine shop was filled with old machines, fitted with modern measuring devices, often with engineers operating on the floor. The Swedish machine shop was filled with ultra-modern machines fitted with the most up-to-date equipment, run by emigrant workers. Nissan Diesel produce over 100,000 units per year, Scania produce 35,000 units per year.

The Japanese focus on maintenance is based on a desire to produce what is needed. This requires 100 per cent availability of plant to perform at a given rate, to meet scheduling requirements. This demands the involvement of operators, maintenance staff, planners, engineers and management across the complete company. In Japan this also involves the close co-operation of machine supply companies with the end users. They are applying the basics of engineering to solve their problems, together. This point, of people working together, is an essential part of the reason for their success in improving productivity, machines and processes.

In a typical traditional company, too often the operators believe that if a machine breaks down, the fitters will fix it, so it is time for a break. In Japan, if a machine breaks down the operator will be the first to see if they can fix it. If they can't, the fitter will work at it while the operator starts to clean the machine, prepare for restarting it or does some other useful work.

How can companies ensure that their machines operate as we need them to?

- Know what is the normal operating condition for the machine.

- Detect early when something is going wrong.

- Put it right again — this often involves improving the basic process.

Once again this appears as a simple solution, the only saving grace is that *it works*. But how? And how can companies implement it? To answer that, we need to look at what is all too typical in our factories.

Unfortunately, in many traditional companies we see factories that often have:

- Dirty machines

- Unorganised work areas

- Spills and clutter around the machine

- No tools present

- Poor or no operating instructions

- Few if any add-ons developed in the factory

- Leaking oil

- Noisy environment

- Vibrating machines

- Broken or non-existent lighting, making work hard to examine closely.

In many traditional companies operators are frequently:

- Not bothered by the dirt of their machines
- Uninterested in cleaning up spills of material or product
- Not trained to understand the operation of their machines
- Not particularly worried if the machine stops — it's break time
- Not particularly aware of the machine's performance over time
- Not trained for the basic maintenance tasks.

In many traditional companies maintenance crews are typically:

- Black magic merchants. They will fix the machine but not tell the operator — or the engineer — what was wrong and what they did to fix it
- Interested in getting the machine going again — quickly — every time it stops
- Uninterested in finding the root cause of the problem
- Very protective of their knowledge and reluctant to admit their inadequacies
- Used to fire-fighting
- Keen to learn and develop if the engineer or manager can overcome the initial reticence and embarrassment.

Most managers understand the term "fire-fighting". In a maintenance context this is a particularly appropriate term. Many engineers spend their days putting out fires. The core ethos of a TPM programme, however, is to leave the ground so wet after putting out the fire that another can't possibly start there again!

PROBLEM SOLVING

That's the machines, the operators and the maintenance staff, but what about the types of problems that face the team of people?

Once again, extensive study all around the globe has led to an understanding of the types of problems encountered in factories, which can be classified as Suddenly Exposed and Chronic problems, as illustrated in Figure 2.17. While both types of problems have been dealt with over the years in a traditional way, the World Class operation tends to go a step further to identify the root cause of the problems and develop specific responses to reducing downtime associated with them.

FIGURE 2.17: TYPES OF PROBLEMS

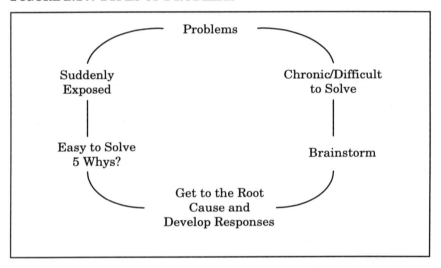

Suddenly Exposed Problems

The main dilemma with this type of problem is that they are too easily fixed. In other words, we don't spend enough time thinking about them, but just fix them. This is the wrong attitude. The root cause of the problem needs to be identified and eradicated. Once again, Toyota are at the forefront of a solution for this problem. Taichi Ohno used the basic question "why?", asked five times over, to find the true cause of problems. This use of "why" or "how" can have a powerful impact on a problem. Once the problem has been tracked to its source, systems improvements need to be made to ensure that it can't happen again.

Chronic Problems

Chronic problems are generally more difficult to handle. They usually involve a number of factors and experience has shown that they are usually related to a lack of understanding and control of a company's process. Quite often both machines and processes are reasonably complicated, with a large amount of information being held by a number of different staff members, but with no single person able to solve all the problems encountered. To overcome this problem the use of brainstorming by problem-solving teams is recommended.

Developing Responses

Basic analysis of the machine or process is required in order to find the root causes of problems. Quite often you will find large amounts of quality reports and fault recording information located at a particular machine. Very seldom is this information collected and analysed and this is the first step in getting a process under control.

Fault Recording Sheets

The simplest form of fault recording system is the diary. Operators can enter such information as production achieved, faults and problems encountered, solutions found, materials used and scrap rates in a simple way. Engineering staff traditionally retrieve this information and use it as a basis for process development. In a WCM environment the problem-solving team includes the operator, maintenance, engineering and management staff, with the quality staff also taking part.

Figure 2.18 demonstrates the idea of recording faults with a machine or process. The creation of an effective WCM Total Productive Maintenance programme must start with the data already available within and from an operation. If the previously mentioned quality and production records are available, an analysis of the faults recorded over a period of time can be carried out. This will typically be performed over a period of two to three months initially. The main types of faults encountered over this period can be collated and identified. This first stage of the analysis should be carried out across the full range of machines in the plant and should achieve a number of important objectives:

1. Identifies main fault types

2. Identifies number of occurrences of faults

3. Identifies amount of lost production due to faults

4. Identifies main problem machines

5. Assists prioritisation of development jobs.

FIGURE 2.18: FAULT RECORDING/ANALYSIS SHEET

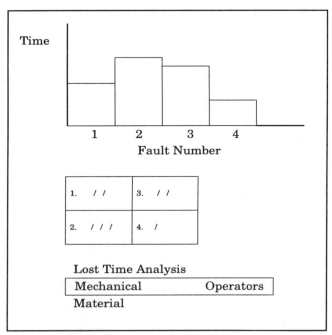

If a machine is complex or large, the use of a sketch can be useful to identify the area where faults occur. Such sketches can be understood and easily used by non-technical production staff.

Information recorded on the fault sheets is readily classified according to the primary reason for lost production, under the three main headings of Machine, Material and Operators.

The machine-related faults can now be analysed to determine which are simple one-off type of problems and which are chronic problems. In the brainstorming sessions with the problem-solving team these faults and problems can be prioritised and addressed. A useful form of prioritisation is in terms of lost time, as well as frequency of problem. The example shown in Table 2.2 demonstrates the idea.

In this example the burnt motor happens only once. This would be felt to be a simple problem with a certain course of action taken to minimise the amount of time lost, possibly the purchase of a replacement motor, possibly upgrading the motor or improving the cooling. The jam in the press and the other faults happen frequently. They may be caused by bad materials, bad operation, bad tooling or any one of a number of other factors. At this point the team needs to try and identify these problems and take action to prevent them recurring. Such actions might include training for operators, developments in tooling and improved lubrication. Such specific solutions need to be identified by the problem-solving team for each operational problem.

TABLE 2.2: ANALYSIS OF MACHINE-RELATED FAULTS

Fault	Frequency	Time Lost / Occurrence	Total Time Lost
Jam in press	27	3	71
Burnt motor	1	165	165
Low air pressure	15	2	30
Head gap	30	1	30

A key point to be understood in this whole process is that operational problems occur on the production floor, they don't occur in an office and generally can't be solved by people who do not understand the process in detail. Office-based support experts can help find the solution, analysis of data can be done in an office but the problems are generally solved in a very practical way, by looking directly at the problem itself, where it happens, when it happens.

The actions of the problem-solving teams will reduce the likelihood of problems recurring, and problems that cannot be removed will be made simpler to fix. Either way, the production time lost due to faults and machine problems will be reduced. Well known and understood remedies will be identified and understood by both operators and maintenance staff. Such fixes will become second nature and eventually they will be designed out of the process completely.

BACK TO BASICS

A number of key issues are involved with Total Productive Maintenance. The most basic of these are:

1. Machines and work areas should be *clean.*

2. Work areas need to be *organised.*

3. Operators need to *care* for their machine. They need to watch, listen, feel and smell their machines to help them identify problems early.

4. The process needs to be *understood* so that actions taken have a positive effect on results.

5. Operators should tighten bolts, maintain oil levels, clean filters — do the *basic maintenance* procedures to help them come to a closer understanding of their machines.

6. Operators need to understand and perform the basic adjustments and maintenance of their machines, and the technical staff need to know how to operate them, in order to enjoy the benefits of *cross-fertilisation* of ability and knowledge.

Cleaning

The simple act of cleaning a machine makes the cleaner more aware of it. This allows the operator to see problems as they develop and to understand the machine better. As the machine inspection system develops, the operator becomes more involved with the machine which helps them notice problems before they cause loss of production time. In any case, it generally feels better to work in a clean environment — a worker has a greater feeling of self-worth when working in a factory that is worth cleaning. A clean area is also generally a safer area to work in as trip and slip hazards are removed. Oil and materials spills should be cleaned up when they occur. Good housekeeping can do a great deal for the overall appearance of a factory and the general level of morale of the staff that work there.

Organising

"A place for everything and everything in its place" is an old adage but an appropriate one. The armies of the world have spent years putting this one into place.

A specific location should be identified for materials, finished goods and parts. Machine-specific tools should be located beside the machine in a tool-holding arrangement. This simple arrangement lets you know immediately when a tool is missing. If it's missing and something goes wrong with the machine, then the time to get the machine going again is inevitably longer. The same applies to raw materials and finished goods. The use of KANBAN squares for both raw materials and finished goods defines locations for both. This ensures that materials can be ready for production with the minimum of supervisory control. A simple look at the KANBAN area will tell everybody if there is sufficient to cover production or if production should continue, thereby avoiding over-production.

Caring

Operators need to care for their machines. Managers need to care for their processes. It is an unusual word to use in an industrial context but it is a legitimate one. When a machine stops, people need to care enough to get it going again, quickly and well. They need to develop this "caring" to the point where they are pre-empting a failure, that is, making adjustments early enough to avoid one. This caring should be nurtured to the point where the operators do the basic maintenance tasks such as cleaning blades, checking drives, greasing and oiling and a myriad of other basic tasks. They need to be trained in this work by the core maintenance staff, who will continue to provide expert support and development capability to the operation.

Understanding

The basic process needs to be understood. A manufacturer has to understand how to get a process back into control if and when problems occur. This sounds obvious, but experience has shown that often companies only *think* they understand their processes.

The level of understanding of the process needs to be spread evenly among all the staff, not just kept with the chief engineer or operators or fitters.

Basic Maintenance

It is essential to carry out basic maintenance. A machine manufacturer will often give you a list of checks to be done on the machine. Quite often these lists can be added to or reduced given sufficient in-house expertise. But the basics need to be done — check bolt tightness, check drive belts and chain tensions, keep oil levels at the correct point, check air pressures, drain water from oil, air and steam lines, and so on. The list is a long one, but quite often the effect of carrying out even the simplest of maintenance procedures is astounding. When checks are being done regularly people come to a better understanding of what the normal conditions should be, which immediately alerts them if the condition of a machine becomes abnormal. Once again, this can help prevent problems from occurring.

Cross-Fertilisation

Just as the operators need to know, understand and do some routine maintenance on a machine, the technical and support staffs need to be able to run machines. This crossing over of frontiers can have very positive effects. Machines can be kept running during breaks. Problems can be avoided as operators catch them before they happen. An adjustment made before a machine breaks usually takes 20 to 30 times less time than does the repair after it breaks.

CONCLUSION

The Just-in-Time concept as part of a World Class Manufacturing programme deals in a very practical way with the operational side of a business. By identifying and eradicating, or at least significantly reducing wastes a company can be more competitive without adding to costs, and be more responsive to customer needs without recourse to larger than necessary stocks.

The basic idea behind JIT implementation is to simplify operations and processes to the point where just what is required is actually done — with no waste! The application of JIT concepts into the maintenance and materials sides of an operation tends to bring previously disparate elements of an operation together. The following chapters on Total Quality Management and Employee Involvement will develop these concepts further and show how the three elements of a World Class Manufacturing programme come together.

Chapter 3

Total Quality Management

ISO 9000

The 1980s witnessed the awakening of industry generally, and western industry in particular, to quality and quality systems. The International Standards Organisation (ISO) introduced a set of quality standards, the most recent in 1987, generically called ISO 9000. The standards consisted of a number of clauses dealing with the core elements of a basic quality system. UK and Irish industry particularly embraced the ISO standards, with many thousands of companies installing ISO systems and attaining accreditation to ISO 9000.

The ISO suite of standards have been upgraded with full co-operation between the international and European standards organisations. The standards are to be known now as:

- ISO 9000-1 (formerly ISO 9000)

- ISO 9001

- ISO 9002

- ISO 9003

- ISO 9004-1.

The first standard, ISO 9000-1, is basically an introductory standard, with the standards themselves starting with ISO 9001 which relates to companies designing, building, supplying and servicing products through to ISO 9004-1 for companies servicing products.

The main changes in the standards concentrate on removing ambiguities and uncertainties, plus there is an additional emphasis on such areas as preventative actions, preservation (of products in storage), servicing and the use of statistical techniques. The total list of headings for the standard are presented in Figure 3.1.

The main difference between ISO 9003 and ISO 9001/9002 is that the requirements are less onerous to comply with for ISO 9003.

FIGURE 3.1: ISO HEADINGS

4.1	Management Responsibility
4.2	Quality System
4.3	Contract Review
4.4	Design Control
4.5	Document and Data Control
4.6	Purchasing
4.7	Control of Customer Supplied Product
4.8	Product Identification and Traceability
4.9	Process Control
4.10	Inspection and Testing
4.11	Control of Inspection, Measuring and Test Equipment
4.12	Inspection and Test Status
4.13	Control of Non-Conforming Product
4.14	Corrective and Preventive Action
4.15	Handling, Storage, Packaging, Preservation and Delivery
4.16	Control of Quality Records
4.17	Internal Quality Audits
4.18	Training
4.19	Servicing
4.20	Statistical Techniques

The original ISO 9000 standards focused very much on a formalised approach to quality management. Procedures were framed to capture what a company was doing in its business. There was no real attempt made to see if what the company was doing were the right things, only that the staff were doing what was written in the procedures. There was no onus on the company to meet independent standards of quality; if the local management team said their product quality was OK, then it was OK for them and their customers. The ISO system was generally perceived as being relatively static.

The changes made to the developed standard have been designed to make the standard more adaptable and more active in improving a company's overall quality of products. The new standard is designed to use the quality system in an active way.

The widespread acceptance of the ISO 9000 system in industry has been of considerable benefit. Companies have taken on board the original system and have at least examined their operations as they wrote the procedures and put the quality system in place. This does not mean that they have become "quality" companies, only that they have a quality system in place. The positive aspect of this work has meant that a large number of managers have looked at and considered their procedures. Many improvements have been made by the companies, but equally often the quality systems have only proceduralised their existing inefficient processes.

Another important aspect which has evolved from the ISO work carried out by these companies is their awareness of the possibility to *improve* their procedures. Indeed, many managers are now actively looking for the *next* way to help them develop their companies. This in turn has left the door open to further initiatives, either in specific areas such as additional quality techniques like Statistical Process Control or process development, or in a holistic approach to business development such as World Class Manufacturing.

WASTE IDENTIFICATION AND ELIMINATION

Identifying and eliminating wastes can be described as a key goal for a WCM programme. The correct use of a quality system and quality control techniques can be very useful in this area. The key tools for this waste identification would be (a) pre-production quality records, (b) in-process quality records and (c) sample analysis reports.

Pre-Production Quality Records

Like its name suggests, this record monitors the product before a production run can be regarded as being stable. Generally a number of key parameters are checked, such as length, width, height, thickness etc., but even more importantly the specification sheet is checked to ensure that the correct product is being made! The check is primarily designed to ensure that the required specifications are being met at the start of a production run.

Details of wastage and lost time to get a process up and running within acceptable standards are usefully recorded on this pre-production quality record. Through careful analysis of the records, team-based solution groups, often including the intervention of someone not generally acquainted with that particular process, can help to identify and eliminate wastes.

As an example, a plastics extruding company was suffering from 3 per cent to 5 per cent waste problems with a significant amount related to start-ups and die changes. A number of typical causes for such wastes would be:

- Heating of dies

- Run through until dimensional accuracy is achieved

- Saw length adjustment.

The "normal" sequence of events for starting up the machine was just as above, i.e. get the machine and dies up to temperature, traditionally by running material through the machine, check the dimensional accuracy of the product for widths, heights, thickness etc. and then check the cut off lengths on the saw. A suggested pre-production check sheet for this operation is presented in Figure 3.2. The suggested sequence for starting the machine would thus be:

1. Pre-heat dies before fitting to the machine

2. Start run through die

3. Get cut-off lengths correct as soon as possible

4. Complete run through of material until full dimensional accuracy has been achieved.

FIGURE 3.2: PRE-PRODUCTION CHECK SHEET

Operator: Shift: Date:
Pre-Heat Die:

Sample No.	Height	Width	Length
1			
2			
3			
4			
5			

Checked:

Clear to run:

These appear to be quite simple developments — and they are — but they all have a good grounding in common sense and will have a visible, worthwhile impact on start-up efficiency. It is important to look critically at what we do, often with a new "eye" to the situation. Advances in measurement equipment have made it both possible and relatively inexpensive for high quality, easy-to-read digital readout measuring tools to be used by machine operators. This simple development allows basically unskilled operators to measure to a level of accuracy that was previously reserved to skilled fitters and engineers who had been trained extensively. This single ability allows more operators to take a greater role in monitoring the quality of parts they produce.

In-Process Inspection

Tests, samples and checks need to be made routinely as a production run continues. These are generally referred to as in process inspection. Most products have more than one attribute and this poses the first problem to management: what do we measure to ensure that what we are making is up to the required standard? Typically there is a key attribute, a weight, a measurement that is regarded as being essential to ensure the quality of that product, and it is this attribute that is the focus for the in-process inspection. There are many ways of recording this information in either tabular or graphical formats. Two methods used widely by companies would be run charts and control charts, both of which are addressed in some detail below.

Run Charts

Run charts are possibly the simplest way to graphically present data over a period of time. The chart allows us to visualise data, with a minimum of difficulty. The run chart can be developed to record the data in number format, which can then be further analysed to give statistical information such as averages etc. A basic run chart is presented in Figure 3.3. In this chart measurements are taken 7 times during the production run and are recorded on the sheet. The readings observed are then plotted on the chart to give a visual representation of the way the measurements varied during the production.

FIGURE 3.3: RUN CHART

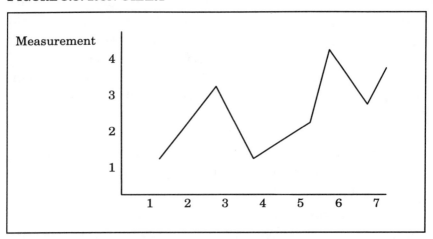

The run chart is usually used as a starting point for detailed quality analysis work and can be a very powerful tool as it is easily understood by the majority of operators. The run chart is very simple in this form but it starts the process of understanding variation. It allows the operator to see easily and quickly just how much variation is present in his system.

Control Charts

The control chart is basically a developed run chart which has had statistical controls added to it. The control chart usually appears with three lines marked on it:

1. The Average

2. The Upper Control Limit (UCL)

3. The Lower Control Limit (LCL).

The control chart can be used to see if a machine or process is running under statistical control, i.e. if there is a high degree of inherent stability in the process to ensure consistency of output. This does not mean that the process will necessarily produce "good" products, but at least it will be consistent. The control chart will be dealt with in greater detail later in this chapter where the formulae will be presented to enable you to calculate the upper and lower limits as presented in Figure 3.4.

FIGURE 3.4: CONTROL CHART

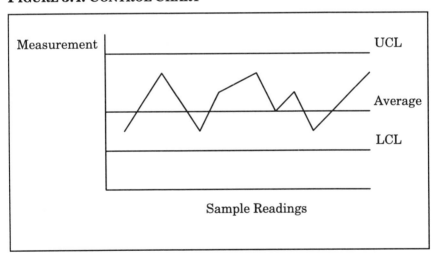

Sample Analysis Reports

The normal, in-process inspection records will generally only deal with a single attribute of the product or process. We know there are usually more than one attribute for a product, however, and this is acknowledged by the use of sample analysis reports. These reports generally look at the key attributes of a product and therefore demand a much higher amount of measurement work than the normal in-process inspections.

Typically, a small number of parts would be measured to this degree from a given production shift. The actual number and the frequency depends on the company itself and the customer requirements. As confidence builds in the capability of a machine or process to consistently perform within specification, there is usually the possibility of reducing detailed examination and instead using spot checks. Many quality sampling procedures have been derived to cover this type of situation, most notably the American Department of Defense's MIL standard. A basic sample analysis report is presented in Figure 3.5. In this diagram the main attributes of the product to be checked were determined as the height, width, length and thickness. The measurements would be carried out on eight parts from the production run to create the completed sample analysis report.

FIGURE 3.5: SAMPLE ANALYSIS REPORT

SAMPLE ANALYSIS REPORT				
Sample No.	Height	Width	Length	Thickness
1	2.3	7	12	0.4
2	2.3	6	13	0.4
3	2.5	7	12	0.3
4	2.4	5	11	0.4
5	2.3	6	13	0.3
6	2.6	7	12	0.3
7	2.4	6	11	0.4
8	2.5	5	13	0.3

Quality reports can be an invaluable source of information in the quest for waste within a factory. The reports outlined above all relate to the production area but there are other equally important sources of waste information within a factory. The use of reject reports, customer complaints and material reports are extremely useful in this work.

Scrap/Reject Reports

Quality and production staff should analyse rejects regularly. Typically, the quality department would report on this matter weekly; however, it is important that production management are aware of quality problems as they occur and are not waiting for the quality report at the end of the week. The reject bin is a very good place to start looking for process improvements. Figure 3.6 is an example of a typical scrap/reject report.

An analysis of this report should either lead to an improvement in internal processes — i.e. to not bend the cross bars or to develop more reliable suppliers — or at the very least to allow management to see where the problems are.

FIGURE 3.6: SCRAP/REJECT REPORT

Inspector: Date:

Reject No.	Reject Description	Fault	Disposition
407	End Pin	No Screw	Hold
408	Cross Bar	Bent	Rework
409	Cross Stop	Missing	Return

Customer Complaints

The last thing a company wants is to have its customers acting as its quality department. However, in the real world, customers can and do find problems with products. When a customer complains there is every good reason to listen and to listen carefully. One point to remember is that the customer's problem may not always be caused by the failure of your product to perform. Do not just accept blindly that you are wrong, but instead analyse critically the complaints with a view to identifying the source of the problem and take the necessary steps to make sure it doesn't re-occur. Maybe the complaint can assist you as a spur to further developing your process.

Machine Reports

Though many factories use machine reports in a superficial way, if at all, they can be invaluable in identifying wastes. A simple report, which only records the day, the shift and the production output achieved is presented in Figure 3.7a. Such a simple report may be all that's required in a perfect world, where machines, operators and processes are running well and under full control.

In the real world, however, management is getting very little help to identify problems from such a report. An improvement on this report would start to address the issue of machine downtime, which is presented in Figure 3.7b. This is an improvement on the first report but is still not very helpful in identifying where and why output has been lost. A more useful form of the report would be as shown below in Figure 3.7c. The report records the same information as above, i.e. day, shift, production achieved and machine downtime, but has also been expanded to provide detailed analysis as to *when* production was achieved and lost as production output is not just recorded as a shift figure but by the hour.

FIGURE 3.7A: MACHINE REPORT

Day: Monday
Shift: 7.00 - 3.00
Production achieved: 3,257

FIGURE 3.7B: MACHINE REPORT

Day: Monday
Shift: 7.00 - 3.00
Production achieved: 3,257
M/C Downtime: 35 minutes

FIGURE 3.7C: MACHINE REPORT

Day: Monday

Shift: 7.00 - 3.00

Production HOURS   1 2   3   4   5   6   7   8

achieved: 400 420 380 410 440 300 278 400

M/C Downtime: 35 minutes

Comment: Problem with machine jams hr. 7. Bad materials.

This report now allows management to get an understanding of the process. They can see if the machine is running consistently or erratically and to identify reasons for lost production. These performance figures can be easily charted to improve their impact.

PROCESS AND PRODUCT DEVELOPMENT

A criticism levelled at the original ISO 9000 standards was that they didn't do enough to actively improve the quality of a product or service. The ISO could be applied equally well to a good or bad product. The new version of the standards is attempting to address this shortcoming.

In a WCM environment quality and performance charts and records have little or no value in themselves; they are valued for their ability to help management understand and improve their processes and products. The information contained in the charts and reports is what is important. A development of the machine report, for example, is the machine fault register. Frequently created by the production staff with input from the maintenance staff, this is possibly the most effective tool in the engineer's toolbox.

One of the main difficulties encountered when trying to develop machines and processes is identifying which parts of a machine are the real problem areas. The Pareto chart, shown in Figure 3.8, represents a useful way of identifying the problem areas of a given process or machine. The areas of the machine are identified as area 1 to 5. The number of faults occurring in each of the areas is then recorded. The number of faults are then sorted in descending order and plotted. This method offers a simple, first phase analysis. As staff members become more familiar with the concepts of WCM, the level of sophistication of the tools used are increased.

FIGURE 3.8: PARETO CHARTS

Area	No. of Faults
1	7
2	3
3	8
4	1
5	2

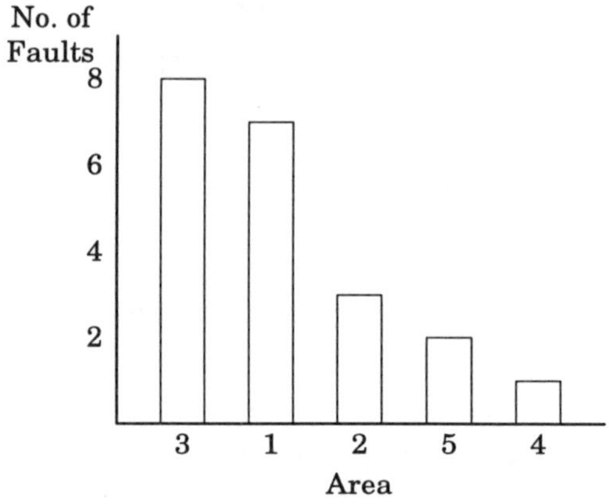

The object of such reports is to identify the causes for the lost downtime identified in the machine report. The Pareto chart helps management identify the faults most affecting production. These obviously form the basis for a process improvement programme. The control charting systems already identified can also be used very effectively in the overall process development procedure.

Process and Layout Development

The WCM principle is based on process development, and it is here that all the tools and techniques come together. The quality system records, the maintenance system records, employee involvement in development teams and overall management commitment to improvement are all key essentials to achieve product and process improvements at a world class level.

The core starting point for improving products and processes lies once again at the micro level. We have talked already about company-wide flow charting and process flowing, now we need to look at the process at the most basic level.

The following diagrams show an actual developmental programme in an Irish moulding company. Figures 3.9 to Figure 3.11 show the evolutionary stages in a process development at the layout level. Figure 3.12 presents the process in words, outlining the steps in each of the above three physical developments.

FIGURE 3.9: LAYOUT #1

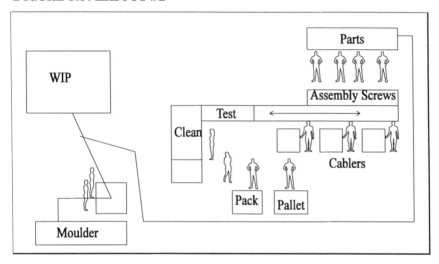

Figure 3.9 shows the original layout, before applying some of the WCM principles. A large floor space (approximately 20 per cent) of factory floor was given over to WIP storage. The flow lines for the products through the process show a lot of to-and-fro movements with very many double handling operations.

Figure 3.10 shows the interim layout. The physical implications for such a move resulted in a 25 per cent saving in overall floor space. Quality has improved as parts are assembled and finished as they are moulded. The process still contained some degree of waste. The WIP area was now reduced to 2 hours' worth of moulding as opposed to 15 days' worth, but there was still waste of storage and handling of the product as it cooled. Two operators were employed full-time in screwing two parts of the mould together, a difficult and fiddly job.

FIGURE 3.10: LAYOUT #2

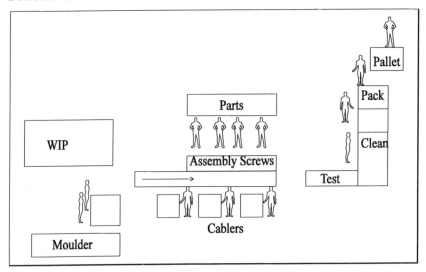

Figure 3.11 shows the present layout. The process has been re-fined further through the acquisition of a heat staking machine which avoids the necessity for screwing parts together. The extended cooling time has been removed from the process. Operators are now transferred to final assembly of units leading to a significant increase in output from the process. Cleaners are no longer required on the line as removing the multiple handling of the product has stopped the products getting dirty in the first place.

Further developments for the process are being examined in the areas of box erection as well as the removal of the conveyor from the operation. This example shows how development of a process is a continuing function. Management used the process flow charting technique to identify possible areas for improvement. Wastes were identified and removed. The process flow charts for these stage of development are presented in Figure 3.12.

This example shows how a process can be developed by using simplicity as a guiding principle, by removing waste actions and where appropriate, through the acquisition of key items of plant.

FIGURE 3.11: LAYOUT #3

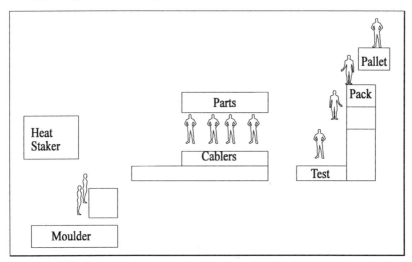

FIGURE 3.12: PROCESS DIAGRAM — STAGES 1 TO 3

Stage 1		Stage 2		Stage 3	
Mould parts	0	Mould parts	0	Mould parts	0
Pre-assy 1	0	Pre-assy 1	0	Pre-assy 1	0
Pre-assy 2	0	Pre-assy 2	0	Pre-assy 2	0
Transfer to WIP	⇓	Transfer to cool	⇓	Heat staker	0
Storage	∇	Storage	∇	Conveyor	⇓
Trans to assy line	⇓	Feed conveyor	0	Assemble	0
Storage	∇	Screw	0	Trans to cablers	⇓
Trans to assemblers	⇓	Assemble	0	Cable	0
Screw	0	Trans to cables	⇓	Conveyor	⇓
Assemble	0	Cable	0	Test	◊
Table/store/sort	0	Conveyor	⇓	Pack	
Transfer to cablers	⇓	Test	◊		
Cable	0	Pack			
Conveyor	⇓				
Table/store/sort	0				
Test	◊				
Clean	0				
Pack	0				

0 = Action   ⇓ = Transport   ∇ = Storage   ◊ = Test

EARLY WARNING SYSTEMS

The analyses and charting techniques outlined in Figure 3.12 all take place "after the fact". Analysis is based on the past, it looks at what has happened — faults, problems and successes. But in a WCM environment it is important to start looking ahead, to stop problems before they happen. The results of the above analysis often provide clues to help us "see the future".

It is generally unnecessary for an operator to watch a machine running. In fact, modern machines run so fast it is often impossible to *see* the operation without the aid of a high-speed video camera.

Many machines come fitted with early warning systems. The sight of lights fitted to machines has become quite common now in companies. There are still large numbers of machines that do not have such devices fitted, and in many cases, if they are fitted, they are frequently left inoperative.

The simplest such form of warning system is usually fitted to a supply system to an automated machine. A decoiling machine is often seen fitted with a stand of two or even three lights (see Figure 3.13). These machines can be set to produce a given number of parts with a given length. The early warning system would typically alert the operator to the following situations:

1. Approaching end of material reel

2. Complete batch

3. Machine jammed or other fault.

As the machine can typically run unattended, the light system alerts the operator that some input is required to allow the machine to continue production. Similar systems of warning lights and/or sirens are used where hoppers feed materials to machines. Figure 3.14 shows a warning system for such a feed. When the level goes low, the sensor becomes clear and the light goes on, allowing the operator sufficient time to refill the hopper. These additional fitments allow the operator to do other work while the machine runs.

FIGURE 3.13: DECOILER WITH SENSORS AND LIGHTS

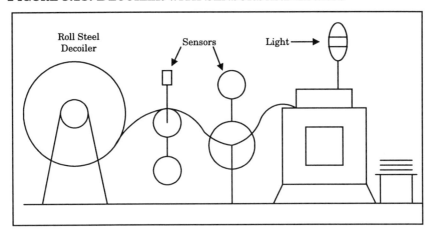

FIGURE 3.14: TANK WITH BASIC ALARM

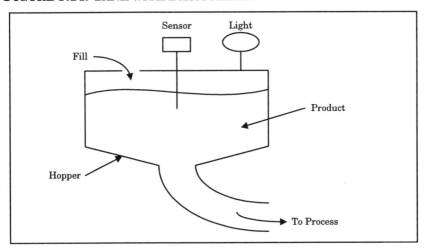

A further development of this system commonly found in the plastics industry is presented in Figure 3.15. A more complicated sensor system or a sensor system with a timer is used to control a feed mechanism from a large bulk system, either a bag or a hopper. Typically the system can operate for prolonged periods without operator intervention while still retaining the ability to call the operator if the system fails. This system allows for good operator utilisation, allowing for large amounts of uninterrupted time to manage the overall process rather than having to watch for material supplies.

FIGURE 3.15: TANK WITH ADVANCED ALARM

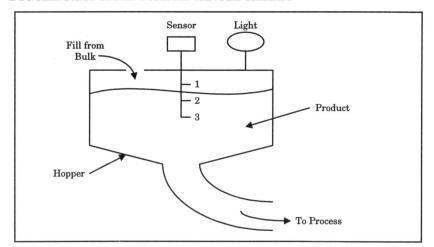

The list of potential sub-automation projects in a factory can often be quite substantial. The problem is usually trying to decide what gives the best return for a given investment.

Andon, a type of early warning system, is the Japanese word for lantern, and though the examples mentioned above referred to lights, these could easily be augmented by buzzers or sirens. The Japanese have extended this lantern concept to include all visual measures. If we take the idea to its most basic level we can understand it better: "If something looks wrong, it *is* wrong!"

When we drive our car we usually keep an eye on the fuel, water gauges and the speedometer. Our fuel and water gauges are generally fitted with green areas (normal operation), and red areas (abnormal operation). The same idea is applicable in many industrial applications. A temperature or pressure gauge can be relatively easily marked with the correct operating conditions.

Holding areas are readily marked to define the quantity of parts to be held there. A production stores area can be laid out to allow the store manager to check easily if they have sufficient parts to complete the orders on hand. If there is an empty slot, they are short a part.

Similarly, on the production floor, the use of a KANBAN square can alert a supervisor or operator to a problem at a given station; if there is a build-up of WIP or product at a given area then there is obviously some problem with the process, a problem that needs to be addressed.

One of the biggest time wasters in many plants is the time spent looking for tools. Only the best fitters ever seem to have the right tools at the right place and time. The use of simple tool boards located strategically can help in this area. If a tool is missing it is obvious and it can generally be found and returned *before* it is needed in an emergency situation. Figure 3.16 shows a device called a Shadow Board. This board is typically painted with the outlines of the tools used at a machine or in a given area.

FIGURE 3.16: SHADOW BOARD

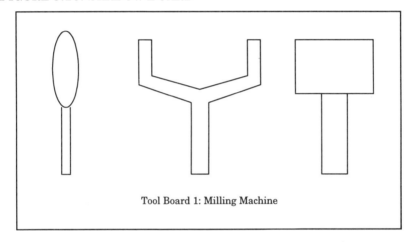

Tool Board 1: Milling Machine

Basic housekeeping can have a significant impact on how a factory operates. A clean, well-maintained workplace generally operates well. A tip heap generally does not. By imposing a sense of order and purpose, management can frequently improve performance and productivity for very little investment.

The use of charts to assist in process development was mentioned earlier. These charts can also have a useful affect on staff performance. By letting your people know what they are achieving they will often want to improve further. Most workers like to know that what they do is valued, that their work is worthwhile; the use of such charts lets them see the result of their efforts.

Typical charts used most frequently would be (a) Production Volumes, (b) Quality Reports — Defects, and (c) Daily Control

Boards, as shown in Figure 3.17. Don't be afraid to show the required production levels for the plant. These should be known and understood by everybody in the factory.

FIGURE 3.17: TYPICAL PRODUCTION CHARTS

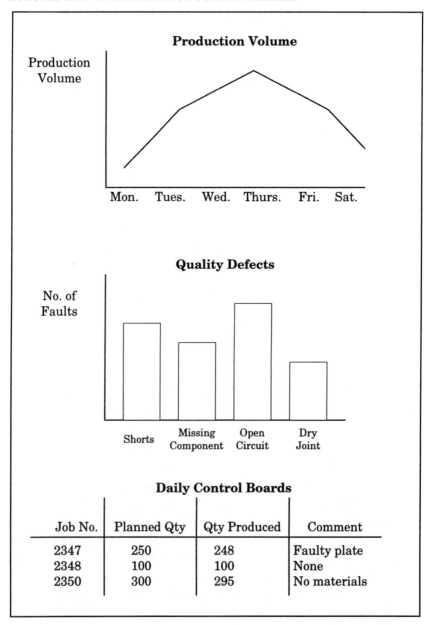

SKILLS REGISTER

The WCM factory demands quite a lot from its staff. There is frequently a requirement to train staff to develop their skills to allow them to perform additional work outside their traditional skills base. The skills register is a relatively easily used and easily understood visual representation of the skills levels of a factories staff. A sample of a skills register is shown in Figure 3.18.

FIGURE 3.18: SKILLS REGISTER

SKILLS REGISTER				
Date:	Job Area			
Name	Job Details			

Staff members names are entered on the left hand side with the major skills areas shown across the top section. A more detailed breakdown of job skills is entered under each of the major skills areas. Under each of these job skills each staff member has a four box square. The number of segments of this box are filled in to represent the ability/skill of the staff member in that particular job skill. A standard interpretation for the skills levels is:

- No boxes — Untrained, unskilled
- One box — Basic introduction given
- Two boxes — Able, under supervision
- Three boxes — Capable, with slight supervision
- Four boxes — Totally capable

The production manager, often with supervisors, completes the form. The completed register can then form the basis for a staff development and training programme. It is very easy to see where specific skills are in short supply and relatively easy to use the register to plan for closing these gaps. When management are happy with their use of the register it can often be hung on the notice board. The timing of this is definitely one for local management to decide.

The production floor notice board can also be used to display awards and commendations for particular staff achievements and suggestions. This again depends on the local culture, but in general, people like to be complimented for assistance they have given to the attainment of a common goal.

QUALITY FUNCTION DEPLOYMENT

The basic understanding of Quality Function Deployment (QFD) is: "Getting the quality mentality to where it matters most — into your people!"

This idea means that workers, designers and management come to understand that quality depends on people doing their job properly. That quality is not "put in" to a product by the quality assurance staff.

The designers, purchasers, workers and managers of a company have the primary responsibility to add their value to a product in a way that enhances the product. All employees of a company need to understand this and act in this way.

The international understanding of QFD is a little more complicated. QFD is used to describe a number of tools and methodologies used to ensure the company designs products that accurately match customer requirements. They are based on matrices and are generally found to be applicable in larger companies which have the capacity to handle such advanced techniques. They can be briefly summarised as described below.

Teams work to try to identify what the customer's needs are and how they can be satisfied. The team should include representatives from the key departments such as sales and marketing, finance, production and purchasing. The key to the process is to actively try to determine what both the customer and the end user actually want from a product.

The team will then strive to find ways to supply these requirements. The process relies on numerical processing, frequently applying weighting factors to particular criteria, and is generally not applicable in an SME-type of operation as the work required to complete such an analysis is costly and complicated.

USEFUL QUALITY TOOLS

The following selection of quality tools have proved to be both effective and simple to use in typical SMEs. A number of other tools exist and are available for specific problems, but these have almost universal applicability:

- Flow Charts
- Check Sheets
- Pareto Charts
- Run Charts
- Histogram
- Control Charts
- Process Capability.

Flow Charts

The flow chart represents the steps in a process and can be a powerful tool when used aggressively. Most quality systems contain flow charts for key processes which usually relate to the theoretical steps involved in making a product or performing a service. This is illustrated in Figure 3.19.

FIGURE 3.19: PROCESS FLOW CHART

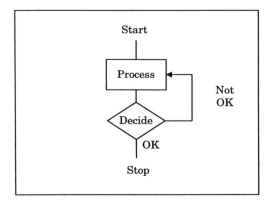

One of the difficulties when using flow charts is getting yourself to look closely enough at the process to see the full detail. Quite often we take it for granted that we "know" the process. This is an often fatal mistake. A example of this is presented in Figure 3.20. The "known process" resulted from discussions with the management. The "actual process" resulted from a detailed analysis of the actual process on the shop floor.

FIGURE 3.20: "KNOWN" VS. ACTUAL PROCESS

Known Process	Actual Process
Put on eyelets	Put on eyelets
Rivet	Bend 2 eyelets down
Bootlace	Bend 1 eyelet down
	Place pin
	Place base and clamp
	Locate wires on rivets
	Rivet and move * 3
	Return to capping stn
	Apply cap and press home
	Remove clamps
	Remove unit
	Bootlace

Note that even in the flows outlined above there is no mention made of pick and place actions. Machinery was relocated as a first step in developing the process, removing an amount of waste associated with a poor layout.

The key point is to identify accurately what actually happens in a given process, compare it with the theoretical process and see where improvements can be made.

Check Sheets

One of the biggest problems we come across when you start trying to improve a process is separating "what people think is happening" from "what is *actually* happening". It is essentially important to separate fact from opinion. If we take action to fix an opinion we may well end up further from the real solution and actually make matters worse.

The check sheet is a very simple, but powerful, tool when it comes to separating fact from opinion. Basically, it records how many times something happens. To use a check sheet you need to (a) define what is being recorded, (b) over what time period, (c) make it easy to record data and (d) be honest when it comes to recording.

One of the regular pitfalls that companies fall into is making initial check sheets too complicated or demanding operators record too much detail. Once again, this is where simplicity comes into its own. As an example, let's look at an engineering company producing precision components.

The company recorded faults per shift in a shift log. The supervisor would examine the log each morning but there was relatively little use made of the information available from this invaluable source of data. An analysis of the log over a three-month period showed the faults presented below in Figure 3.21.

FIGURE 3.21: PRODUCTION LOG ANALYSIS

Faults	*% of Total Number of Occurrences*
Straightener	6
Footgap	28
Misfeed	20
Angle	7
Cut-Off	22
Length	7
Other	2

Armed with this rough analysis, management discussed the problems with the operations staff and decided to create a check sheet to be used on the floor. The check sheet was used to ensure that the real faults in the process were correctly identified. As all the team members knew what the objective was, management was confident the real problems would be identified. The check

sheet used (see Figure 3.22) was prepared and issued to the production staff. A sample of the filled out sheet is presented.

The initial reaction of management to the analysis outlined
above was "let's attack the misfeed, cut-off and footgap problems
immediately". This was a valid response but not necessarily the
best one for the company to follow.

This analysis formed the basis for further study which will be
dealt with under the Pareto Analysis section.

FIGURE 3.22: ENGINEERING COMPANY CHECK SHEET

Engineering Co. Check Sheet
Misfeed ‖‖‖ ‖‖‖
Cut-Off ‖‖‖ ‖‖‖ ‖
Footgap ‖‖‖ ‖‖‖ ‖‖‖
Straightener ‖‖
Burr ‖‖
Angle ‖‖
Length ‖‖
Other ‖

Another example can be taken from our electronics company. After the company had relocated its machinery and developed its
process it still felt there were additional possibilities for improvement. They turned their attention to further fault reduction.
A check sheet, Figure 3.23, was designed for the production area
to help identify the root cause of the 5 per cent of faults still being
identified at the touch-up area.

This basic level of analysis allowed the manufacturing team to
further improve the overall operation. An earlier attempt at fault
recording had resulted in a list of 17 possible faults with the operators responsible for recording individual locations and part
identifications. The system was withdrawn soon after its introduction. The simpler check sheet has been used actively by the
staff.

FIGURE 3.23: ELECTRONICS COMPANY CHECK SHEET

Electronics Co. Check Sheet — touch up
Fault
Dry Joint | | | |
Missing Comp | | | | | | | | | | |
Short | | | | |
Wrong Ins. | | |
Other | |

An essential point to be made in relation to check sheets is that the manufacturing team, either in a cell, a process line or at management level, *must use* the information gathered. If actions are not taken on the information obtained the check sheet itself becomes a *waste*.

Pareto Charts

A development from the basic check sheet can often be a Pareto-type chart. A Pareto chart can be regarded as the representation of the ordered data from a check sheet. In the previous precision engineering example the company recorded the frequency of faults. This data recording was developed to also record the lost time per fault. This data is now presented in Figure 3.24.

FIGURE 3.24: DATA FROM DOWNTIME LOGS

Engineering Co. Faults & Times Lost		
Fault	*No. of times*	*Time Lost*
Misfeed	10	50
Cut-Off	11	20
Footgap	15	52
Straightener	3	90
Burr	4	130
Angle & Length	4	42
Other	1	5

This development of the check sheet data puts a different complexion on the problem. If we now order the faults, as shown in Figure 3.25, we see that the burr problem and the straightener needed to be addressed with a high degree of urgency. Solutions were found using brainstorming techniques, improved housekeeping and general process development including increasing materials and tooling knowledge.

FIGURE 3.25: ORDERED DATA FROM DOWNTIME LOGS

Faults ordered by time	
Burr	130
Straightener	90
Footgap	52
Misfeed	50
Angle & Length	42
Cut-Off	20
Other	5

Run Charts

A run chart is used to display trends over a period of time. By presenting data in a graphical way we can usually detect trends more easily than by just looking at numbers. Common examples of such charts would be production output, customer complaints received, sales made and any one of a host of others.

The charts are particularly useful when it comes to determining if actions being taken have an appreciable affect on the underlying process. Samples of these charts are presented in Figures 3.26 and 3.27.

In the first of these two examples, production output improved by approximately 25 per cent, when the management structure of the firm was re-organised, with better focus on the primary task of making product.

FIGURE 3.26: RUN CHART 1

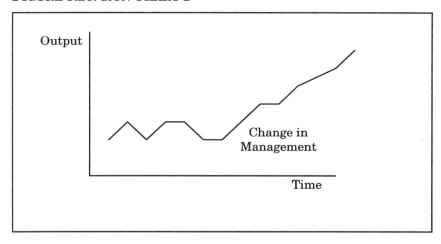

The second example relates to a plant where production was being crammed into the last two weeks of a month, through a typical cash management style of production management. The company moved to a cellular manufacturing production system, improved output, productivity and customer service.

FIGURE 3.27: RUN CHART 2

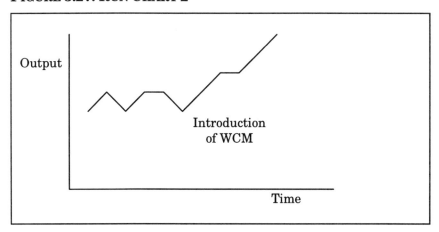

The run charts show such changes in trends quite clearly and can be an effective tool to keep an eye on your key processes, easily.

Histogram

When we start using histograms to analyse data we are fast approaching the area of the quality professional. However, we can often make use of a tool without needing to understand its innermost workings — take the computer as an example.

The histogram represents information using bar graphs. The information is generally the frequency that certain data occur, for a given process. Basic analysis of processes shows that repeated events will vary over time. By measuring and monitoring these variances we can come to a better understanding of how the process is operating. A typical histogram is presented in Figure 3.28.

The parts produced by the process are measured for thickness. The range of thicknesses is between 3.3 and 3.6mm. The vertical axis records the frequency of occurrence of the given thicknesses. The histogram represents this data as a bar graph.

FIGURE 3.28: HISTOGRAM

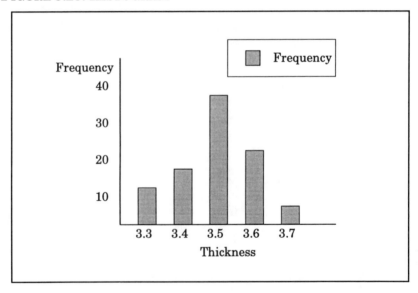

A histogram can be very useful when running a process, as it gives a view of how the process is varying. If in the previous example, the specification for the thickness was set at 3.3 to 3.5 we would see that the process was producing parts on the high side of the specification, with some parts being rejected. By regularly monitoring the process we can see if the process is shifting either high or low. The next two tools will address this matter of control and capability further.

Control Charts

A control chart is a development of the run chart. It uses statistical methods to determine upper control limits and usually also lower control limits. These limits are calculated based on measurements taken from a process running in the normal way. When gathering this data the process should not be adjusted over and above the normal adjustments made during a run (see Figure 3.29).

FIGURE 3.29: CONTROL CHART

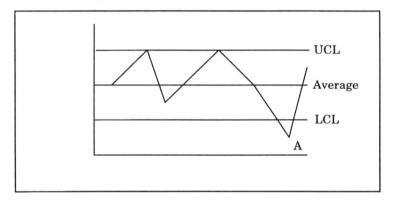

When you determine the upper and lower control limits, run your process to see if you are running within your control limits. This does not mean that you are necessarily making good products, only that you are making *consistent* ones. The control charts can be very useful, however, in tracking down problems and providing facts about a given process. The possibility of opinions clouding the issues is much reduced when a process is running under statistical process control.

By changing and developing the process it can be possible to reduce the gap between the control limits and be sure that the revised process is once again under control. When you graph the results from your process you will see variation between the upper and lower control limits. The statistics of the process tell us that these variations are a natural feature of the process, dependent on the system design, machine type, maintenance schedule etc. Points such as point A shown in Figure 3.29 are as a result of some special cause, such as an operator error, change of materials, unplanned events etc. These special causes need to be identified and removed from the process before it can be said to be in control.

A key point to remember when dealing with control charts is that a process *in control* is not necessarily a good one. It may well still not meet your specifications. Specifications are what you think you need to satisfy requirements — control limits are what the process itself is capable of delivering.

FIGURE 3.30: DEVELOPED CONTROL CHART

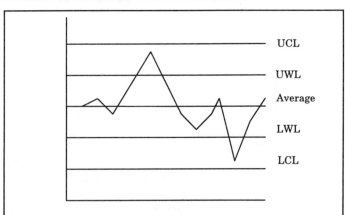

A process is said to be in control when it satisfies a number of criteria. If we take the UWL as the upper warning limit, and LWL as the lower warning limit, we can say the process is in control when:

1. No value lies outside the control limits

2. No more than about 1 in 40 values lies between the warning and control limits

3. No incidences of two consecutive values lying in the same warning zone

4. No runs of more than 6 values which lie either side of the average line

5. No runs of more than 6 values all rising or all falling.

When we have the process under control we can now address the question of capability of the process. We can determine whether or not the process can satisfy our specification requirements.

Process Capability

We have talked so far about processes and how they function themselves. We now need to see how a process can meet external specification requirements. The control chart can tell us when a process is in control, but it can't say whether or not it is a good process.

Capability indices have been developed to allow objective measurement of the ability of a process to meet given specifications. Though somewhat technical, it may be worthwhile to describe the process in detail.

If USL and LSL stand for the Upper and Lower Service Limits, the formulae for determining process capability are:

$$Cp = USL - LSL \div 6\sigma$$

When a process is in control we can estimate σ, the standard process deviation from the control chart:

$$\sigma = \bar{R}/d2$$

Where $\bar{R}$ = Average of subgroup ranges and
d2 = A constant defined by the size of the subgroup samples

Where for	n	d2
	2	1.128
	3	1.693
	4	2.059
	5	2.326
	6	2.534
	7	2.704
	8	2.847
	9	2.97
	10	3.078

Cp is a basic capability measure as it looks at the process in relation to the width of the specifications, it does not relate to how close the process average is to the required target value. A further development of Capability Indices uses Cpl, Cpu and Cpk where:

$$Cpl = \overline{\overline{X}} - LSL \div 3\sigma$$
$$Cpu = USL - \overline{\overline{X}} \div 3\sigma$$
$$Cpk = Min\ (Cpl,\ Cpu)$$

The measure of Cpk is taken as the process capability, if Cpl = Cpu, the process is centred around the target value.

If Cpk = >1, the process is capable. The greater the Cpk level than one, the more stable and capable the process is with an ever-decreasing capability of the process to make defective product.

CONCLUSION

Total Quality Management is a fundamental element of any WCM programme. We have looked at how quality systems, basic quality tools and some more advanced quality techniques and concepts can be integrated into overall process and product development. If we were to look for one particular distinguishing feature that differentiates traditional quality approaches from that of a World Class company quality approach, it would be that in a World Class company, quality is proactive rather than reactive. World Class companies constantly use quality information and tools to help identify further areas for development, to identify wastes and generally provide much of the base data upon which a successful World Class Manufacturing implementation is founded.

Chapter 4

Employee Involvement

World Class Manufacturing has been described as a process, a series of techniques and tools, a programme. It is in fact a state of mind that uses the above to help companies improve themselves. The management and workforce of a company both need to be fully supportive of WCM if it is to have any chance of success within the company.

As WCM is so broad-ranging it is very difficult for managers and employees to understand the detail of a full WCM implementation, but they do need to understand the basics. They need to understand that WCM involves looking for wastes, working with people to improve the product or service being offered to customers. So far we have focused on the manufacturing side of the business. We will now look at the non-manufacturing aspects to see where WCM principles may be applicable.

APPROPRIATE STRUCTURES

The traditional concept of management led us to our current practice of departmental heads and general departmentalisation of the business functions. The financial controller watches the money, the sales and marketing manager deals with the customers and markets, the purchasing manager deals with the suppliers, the personnel manager deals with the staff and so on. The typical structure, as shown in Figure 4.1, is familiar to us all in the West.

If we think back to the manufacturing area we can see the organisational similarities. Delays and build-ups take place between the departments. Individual managers will optimise their own areas but will give little thought to how they can reduce waste across the company. The great drawback with this arrangement is that very few customers deal with only a single department. Most of the requirements of customers are handled across the structure rather than up and down it.

FIGURE 4.1: TRADITIONAL ORGANISATIONAL CHART

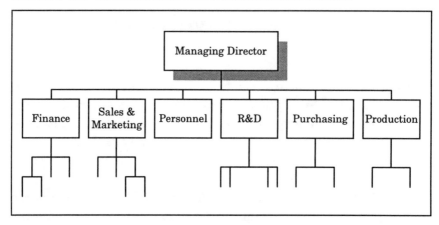

Let's use the flow charting tool (Figure 4.2) to examine how a typical customer's requirements are satisfied by the company.

FIGURE 4.2: PAPER FLOW THROUGH THE ORGANISATION

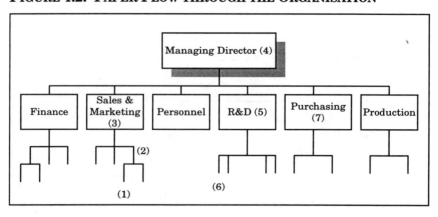

Sales representative 1 makes a client contact for a new product; they discuss the matter with their supervisor, 2, who thinks it's a good prospect and not very different from the existing product range, with reasonable volume and profitability expectations. The supervisor discusses the matter with the marketing and sales manager, 3, who agrees with their supervisor (sorry, area sales manager!). The sales and marketing manager brings the matter up at the next management meeting and the MD, 4, clears the R&D manager, 5, to investigate the matter and report back — directly to the sales manager, to avoid waste of time and effort!!

The R&D manager, 5, goes to the project engineer, 6, who refines the design and all the other things engineers do. They pass their work to the R&D manager, 5, who talks to the purchasing manager, 7, who carries out a procurement exercise and reports to the R&D manager, 5, and the sales and marketing manager, 3, who responds to the MD, 4, who gets the financial controller to examine the implications of the new business on the company's finances. The story continues . . . but you get the picture.

Now, if the business could be organised slightly differently:

FIGURE 4.3: TRADITIONAL VS. WORLD CLASS ORGANISATION

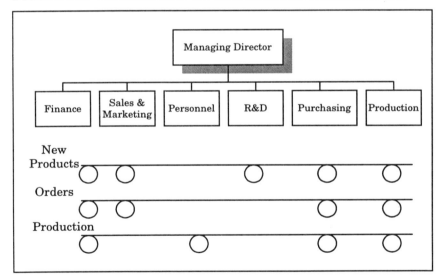

In the new organisational structure the key processes have been identified. The use of flow charting allows us to determine the course of the processes and identify wastes inherent in them. Once these wastes have been identified we have a chance to remove them from the picture.

Figure 4.4 shows a representation of the paper trail in a typical order entry office. The overall lead time is 2 to 3 days with the actual processing time of 1 1/2 hours. This simple analysis shows us there is a possibility to reduce lead time and thereby improving efficiency in the operation. The process was analysed above and this showed that the same information was being entered a number of times, by different people. If the system were computerised, however, it would look very different, as shown in Figure 4.5.

FIGURE 4.4: PAPER TRAIL 1 (DAYS IN PROCESS)

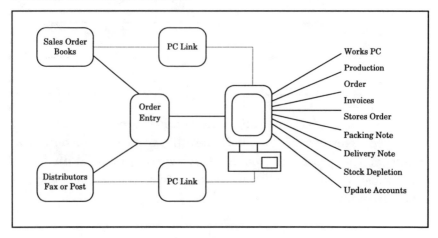

FIGURE 4.5: PAPER TRAIL 2

By using PCs and telecommunications, it becomes possible to link up the salespeople and distributors directly to the company's own computer system. Orders can be entered by the sales and distribution staff, ensuring that the orders placed on the company are what is required by the customer. The overall efficiency of the system is improved significantly.

The use of PC to PC communication allows for the removal of an additional step from the process. Data is entered only once into the system. Previously a significant proportion of lead time was taken up between the placing of the order with the salesperson and the time that the order is actually entered onto the production schedule. The use of PC-based communications helps to remove this waste.

	Lead Time	Work Load
Without modem links	1 day	5 mins
With modem link	—	Admin only

The analysis is based on identifying what actions add value to the transaction, removing those that add no value and automating what is left.

EMPLOYEE INVOLVEMENT

One of the features of a WCM company is that it understands the importance of its people. But this is only normal, we say. The difference in a WCM company is that the people are not restricted to their particular jobs. They are expected to look at any problems facing the company and attempt to fix them. At a particular level this means that people have to *think* as well as do to earn their keep.

The term Employee Involvement has been coined to cover this new awareness. As the name suggests, employees become actively involved in the company. Their contributions to solving problems are sought and they become integrated into the overall problem-solving ability of the company.

The Employee Involvement side of WCM is the *"together"* part of the simple WCM definition, presented at the beginning of the book. This all sounds very American and Japanese but it has been proven to work in European companies, ranging in size from the very small to the very large, from open management styles to basically autocratic ones.

In the open management-style company, people are usually aware that they are valued and that they contribute usefully to the organisation. In the autocratic-type company this is usually not as obvious. The open style companies can generally carry out large elements of a WCM programme without any outside help. They may need guidance through some of the more unusual concepts, and may even need assistance to implement some action programmes, but at least the element of fear is not usually an issue for them.

In the autocratic-type company, however, employees usually exhibit a degree of fear. Rule is by diktat, and there is generally only one way to do things, the *boss's* way. Many of these bosses will never read this book, and many of them will only come to look at WCM when their companies find themselves in a crisis. But it is not too late! If the bosses can come to terms with the fact that their "ill performing team" may be useful after all, then there is hope. The boss has to try and put some faith back into the employees, try to help them to develop their capabilities and help to turn the company around. It isn't easy but it certainly is possible.

These two pen pictures represent the opposite ends of the styles of management found in Western companies. They are extremes but if you look at your company closely you will often find matching examples of these behaviour patterns across the company.

CULTURE

Over the years a lot has been said about the culture of a company or organisation. Leading examples have been put forward such as Apple and IBM, where the culture was supposed to be built into the total fabric of the organisation. We know of similar examples in Europe. But this in-built "in the fabric" culture is not as inherent in the organisation as it is in the leader of the organisation. For example, IBM has been in the wars over the past few years. A new director was appointed to the top and the IBM culture changed dramatically. Many IBM staff now use "warehouse"-type offices. They just come in, plug in their portable computers at an empty space and do their work. Largely gone is the idea of Corporate Headquarters. The culture has changed.

The point is that culture reflects the type and style of leadership of the managing director. Many autocratic leaders work this way because "nobody else cares or will do the job right!". It is essential that such leaders come to understand that most people will care and will do the job right, if they are given the chance, the training and the power to do so. What difference does it make to a company to have one person caring and doing right or to have 20, 50 or 100 people caring and doing right?

TEAM BUILDING

Team building — it sounds so mechanistic, so formal, so remote. We all know what a team is — "a group of people working together for a common goal" — but how do we build a team? At the outset, we must realise that just bringing a group of people together and calling them a team achieves very little, if anything. There needs to be a specific goal or task to concentrate the minds of all concerned and focus them on the attainment of this common goal.

Much has been written about teams, how to put them together, apply techniques and analyses to ensure you get the right mix of skills, personalities, motivations and so on. Once again, this fully logical, mechanistic approach works well where you have large numbers of people, lots of resources at staff support levels, and considerable amounts of time and money. Unfortunately, the majority of SMEs don't have access to these resources. Typically a company has a small number of managers available, a small number of supervisors and a restricted amount of support. So how can teams be created where you don't necessarily have all the required mix of skills and personalities?

The Facilitator

The use of an external facilitator can be the answer to this question. This person needs to get an understanding of your company and your problems and then help you through the stages of putting an effective team together.

The role of the facilitator is a particularly sensitive one. The facilitator needs to be able to move from an aggressive position where they will push the team members to look for solutions to one where they will be totally receptive to ideas and opinions expressed. They need to ensure that all members have the opportunity to make contributions to the discussions and must be able to moderate any heated discussions that may arise.

The role is particularly difficult in that the facilitator is obviously *not* a part of the company and *not* a part of the team, while very much being a part of both. The task of the facilitator is similar to that of a catalyst in a chemical reaction — without their being there, very little if anything would change.

Because of the complexity of the role it is somewhat difficult to outline how it evolves. Here is an attempt at delineating the stages in a typical facilitator role:

- Lead off the team

- Guide through the selection of topics

- Initiate open discussions

- Push the team members to contribute and perform

- Let the team develop its own thrusts

- Gradually devolve power to the team

- Monitor performance of team versus tasks

- Praise and reinforce achievements

- Sit back, drink tea and listen

- Withdraw.

As the team develops, the team members become more confident and they can see how they have actually achieved things they didn't think possible. The facilitator needs to help the team to develop to the stage where they are no longer needed and can withdraw.

In the autocratic environment the facilitator needs to try and protect the fledgling team. They need to block the autocrat, let the boss see how the team is building and effectively alter the basic autocratic tendencies of the boss to the point where they don't destroy the team and its work.

TEAM EVOLUTION

Whether or not you use the services of a facilitator, there are a number of stages in the development of a team that will be observed:

1. Forming

2. Storming

3. Norming

4. Performing.

Forming

When a group of people are brought together to form a team, group dynamics start to come into play. Individuals will tend to wonder where they fit into the team. Why are they there? What is their status relative to other team members? The team generally tends to be relatively polite, carrying on in very much the normal way for their company. It is usually keen to get on and do something, at least in a non-autocratic company. It wants to achieve and generally morale is high. New concepts and ideas can be taken on board and they will tend to learn quickly. In the autocratic company, people are generally slower to say their piece. They will wait to see what the boss has to say, to see how he or she approaches the team and its tasks. This point is very important for the future of the team. If the boss is seen to be supportive of their contributions and their efforts then the team has a future. If not, then the team is starting from a very poor position.

When the tasks for the team have been introduced and decided upon, there is still a lack of ownership of them by the team. They may not fully understand the tasks and generally have not reached to the stage where the tasks can be completed, but they will feel that, yes, they can solve them. Team leadership has probably been arbitrarily assigned to an individual, or is being carried by the senior manager or the facilitator. The team itself has not fully accepted this leader, but has no opportunity or ability to alter the position. The leader will lead discussions, outline the goals and objectives of the team, decide how and when the team will meet and work and generally does most of the talking at this stage.

Storming

The "team" has been in existence for a number of meetings now. Some of the members are beginning to feel annoyed and frustrated at the lack of results, or the way problems are being addressed. They are beginning to *care* about the goals and tasks of the team. This caring often exhibits itself as frustration and anger at the commitment of others to the goals of the team. People become very animated in their discussions.

Often, team members will start to compete for position within the team. This in itself assists the team as the level of enthusiasm is driven up as results are chased. Some of the team members will really start to question their existence within the team, they will feel they are contributing little in comparison to the "young lions" who are striving forward to solve the problems. These team members need to be reinforced at this stage of the team building process. Their contributions need to be commended and appreciated by the leader. An objective analysis of individual team members contributions to the team efforts starts to take place. The young lions begin to appreciate that other members of the team have made useful inputs to achieving the team goals. People begin to understand and appreciate the different abilities, skills, contributions and commitments of other team members to achieving the *common* goals of the team.

The actual tasks outlined for the team will see some successes and some failures. The overall capability of the problem-solving team will not be fully realised yet, with some of the more forceful team members still making excessive contributions to the "solutions". But new ideas and initiatives will come out of the successes and the failures achieved by the team. New approaches to the failed solutions will be proposed and tried.

The team leader's role is critical at this stage of the development. They will have to be both aggressive and conciliatory. They will have to pull suggestions and contributions out of the quiet team members and push back the young lions. They will be generally disliked by both sides. But they will be getting things done! The issues facing the team will be addressed. Actions will be agreed and carried through. The leader will often have to strike a fine balance between criticising and congratulating the team. They need to develop the feelings of self-belief, self-worth and ability within the team. They need to develop the commitment of the team, to the team.

Norming

By now the team has achieved some visible level of success. The team members have got over their initial feelings of inadequacy and have come to accept their fellow team members for what they are and what they have to offer. They are generally comfortable with each other within the team and don't want to raise problems with each other. Their level of interest in the team has risen and they will tend to manage themselves rather than be managed. They will tend to set their own agendas and know what they can and cannot achieve. They will tend to be wary of people outside the group.

The tasks the team set for itself will be coming towards achievement. The previous failed solutions will have been re-examined and probably resolved. The interaction between team players will be much enhanced with real group problem-solving contributing to the solutions. New ideas will be forthcoming for any of the outstanding problem areas.

As the team becomes more comfortable together the team leader needs to draw out discussion and conflict to ensure that real solutions are being reached to the real problems of the company. The leader will tend to move from being a leader to being an equal member of the team. Responsibility for the team's performance rests with the team rather than with the leader, who tends to integrate fully with the team but still reinforces and congratulates achievements.

Performing

The team has now reached the height of its productivity. People are proud to be a part of the team. Others in the organisation want to be a part of the team. They see that the team is achieving results and appreciate this. The team members themselves have come to trust their colleagues and respect their opinions. As a team they can move hard and fast over open difficult ground to solve problems. They are not afraid to have arguments as they know the discussions are being held to find a solution, not to score points off each other.

The initial tasks set for and by the team have been completed. Often, these tasks have been superseded by even more ambitious plans and targets. The team is solving the problems together in an active and aggressive way, happy to report back to the team on tasks completed, happy to take on responsibilities for the team. The team is operating with a minimum of waste and is very protective of itself and its members. Nobody wants to leave.

The team leader no longer exists as a separate entity. They have integrated as a full team member, with responsibility for team performance being equally held by all the team members. The leader will probably retain a steering function to ensure the team does not veer from its course or its standards.

TEAM BUILDING — THE MECHANICS

The team building process has a number of specific steps that are universally applicable:

- Team selection/ Team leadership
- Team objectives
- Team meetings
- Team assignments
- Team dynamics
- Team results and overviews.

Team Selection

Teams are created in a company to solve problems. The successful team needs to have the right members. It must include people with the appropriate knowledge, skills and experience on the team to enable it to find solutions. If a problem is related solely and completely to the production area then there is little point in having a member of the accounting or engineering staff on the team. If the problem extends beyond a specific area of the company then the appropriate staff members *need* to be part of the team. As an example let's look at two teams in a factory:

Team 1
Production has been experiencing some process difficulties relating to the operating procedures on a given machine. They form a team of key operators, maintenance staff and production supervisors to define a standard operating procedure and training programme to address the problem.

Team 2
The same company is having a number of problems with parts shortages. This appears to be a stores and purchasing problem. When an analysis was done on the causes of the shortages (using a check sheet system), it became apparent that part of the problem related to:

- Forecasting accuracy (involving Sales and Marketing)

- Bills of Materials errors (involving Engineering)

- Accounts Payable (involving Finance)

The team formed to address the shortages problem consisted of members from:

- Stores

- Purchasing

- Sales and Marketing

- Engineering

- Finance.

The team resolved the individual issues raised by the shortages problem using this cross-functional team approach.

Team Leadership

Team Leadership is a critical aspect of WCM team problem-solving. The importance of the team leader's role has been stressed throughout this chapter. The team leader needs to know about the practical use of a wide range of tools and techniques such as check sheets, flow diagrams, brainstorming and other tools as well as a practical approach to psychology. The leader needs to be able to direct and assist the team through its tasks. They need to be sensitive to the tasks and to the people, to know when to push and when to listen. Care and attention needs to be given to the selection of the team leader as well as the development of the leader in the aspects of the role they may be weak in.

Team Objectives

The team needs a set of objectives. These must be understood by all the members and be the focus of the team's work. As the team will be in existence over an extended period, it is important that the team does not lose sight of the objectives. It is up to the team leader to remind the team, frequently, why they are there! Often a team will tend to wander off the specific subject and the leader has to remind them of their core objective. This needs to be handled carefully as a slight shift from the core area can often result in a new perspective from which worthwhile solutions can be found for the core issues.

Team Meetings

Team meetings are the place where members can report back to the full team on actions taken, problems encountered and achievements made. The meetings should follow a reasonably defined agenda. At the early stages of team formation this should not be particularly rigid, as the WCM concepts are quite difficult to come to terms with and will require free-flowing discussion between the group. The first few team meetings are therefore very free-flowing, addressing the concepts and answering questions of the team members. As the common understanding of the team develops, a formal agenda becomes more important. This should take the form of concise reporting from individuals on task status, with the body of the meeting given over to further developing suggestions to address outstanding problems.

The meetings can often be used to develop the team members' understanding of particular tools and techniques that are appropriate to help solve the tasks posed by the group objectives. Minutes of the meetings should be prepared and circulated as soon as possible after the meeting, as this ensures the group dynamic is maintained, with all team members being kept up to date with the progress of the team and what is expected of them for the next meeting.

Team Assignments

The WCM philosophy is based on analysis followed by positive action. Teams which meet and do nothing else are a waste of time and effort. The use of team assignments allows the teams to *do things*. Specific actions and plans are created by the team with direct allocation of specific tasks to specific team members. These plans and task assignments should be agreed in open session at the team meetings with the full team being aware of and agreeing the allocation of tasks.

Team Dynamics

The evolution of teams was described in the previous section. The interaction between team members was identified as being critical to the success of a team. The team needs to be a place where creativity is encouraged and assisted. New and innovative approaches to problems need to be brought forward. Here is where the leader's role is critically important, as they have to:

- Create an environment where creativity grows

- Allow everybody to contribute

- Reinforce participation

- Ensure that reality prevails, not opinions

- Help the weak team members to participate

- Hold the strong ones, at least temporarily

- Assist people to listen openly to suggestions

- Foster the group to take responsibility for its actions.

Team Results and Reviews

Teams are created to solve problems. The dangers of acting too quickly on opinion rather than fact have been outlined earlier. The team needs to ensure that it addresses the REAL problems that it has been set. In this context, the review process is necessary. At regular intervals the team needs to check that it is effectively addressing the objectives it was set. It needs to monitor its progress against these objectives. The systematic approach to problem-solving helps the team to monitor its own progress.

- Record data — Measure and record results

- Analyse data — Look for information *from* your data

- Use data —Don't just collect data, use it!

- Act on the results — Use the information to improve the process.

Empowerment

Now that the company has taken on the idea of teams addressing their problems we have to allow the teams to act. We need to empower the teams to change and develop the processes they have been asked to work on. This is often a particularly difficult thing for managers and supervisors to come to terms with. They are used to being the brains of a company; they see, think and decide what action to take. It can be very uncomfortable when workers start thinking for themselves. This is where the process of team building and an awareness of WCM principles come into play. As managers go through the team building process they will come to terms with the abilities and contribution that all the team members can make to solving problems within a company. Then, the question of empowerment does not arise. How can they stop a group of people carrying out actions which they have helped to decide upon?

In the traditional approach to process development, managers, industrial engineers and others would be sent off to "improve the process". They would design new work practices, new layouts and define new improvement programmes. These initiatives would often fail, as workers were not involved in the development, were not committed to the new idea and didn't really care about the process.

In a WCM environment there should be no such split between Them and Us. Process improvements are for the benefit of all and each team member has a role to play in its implementation. In the old way of thinking, people would say "it's not my job to improve things". Now, in a WCM environment, people say "why shouldn't I help improve things".

This small but extremely significant shift in attitude accounts for much of the potential improvements in the processes. While we had looked closely at processes and systems using the tools of WCM we have also looked at the soft side of the development process. We are trying to develop people's caring instincts for their work, possibly re-capturing some of the pride people feel for the work they do.

WCM has been described as a move back to mass craftsmanship, where people are proud of their work. In many cases where WCM programmes have been put in place in companies, cell staff and supervisors have initially been slow to come to terms with the concepts. After three to six months of working in a WCM environment they have become very enthusiastic and keen to show you their latest developments and improvements on the basic ideas put in place at the start of the programme. They have come to take ownership for their own areas and processes.

Management can have difficulties with this idea of staff running their own processes. Supervisors in particular have to develop a different approach to their work. They are no longer people watchers and chasers; they become teachers, assistors and system developers. They help their people to help themselves.

As management have been involved in the development process through the analysis stages, the team building and problem-solving stages, they find it relatively easy to accept that people will want to and can effectively make changes to improve their processes.

QUALITY AWARENESS

The ISO standard system has had a major positive impact on industry over the past 8 to 10 years (see chapter 3). One aspect of the system, however, has had both a positive and a negative effect. The standard requires a company to appoint a quality manager or representative. This has meant that the issue of quality within a company has a champion, but it also has the effect of compartmentalising quality under a single department head.

"Quality Is Everybody's Responsibility!"

This simple statement forms the core of the WCM idea of quality. Quality cannot be added to a product, after the fact, by a quality inspector. No amount of measuring, testing or counting of a product will enhance its basic quality. Quality can only be added to a product or service during the value-added parts of the process. This adding of quality starts with the designers and works its way through the company until the end user receives the product. All staff members need to be aware of this quality chain and to understand that they are an integral part of it and not separate from it.

This awareness of quality and how all company employees are part of the basic quality of a product needs to be introduced to staff by a committed and dedicated management who care about their business and their customers. Training of staff becomes an essential part of getting this message across — training in the basic concepts of WCM, in the use of specific tools and techniques and in developing staff in interpersonal and non-job-related areas. Training needs to be integrated into the overall programme.

Staff members should be introduced to quality tools as they become appropriate to them, as the level of involvement in the improvement process develops, and as the responsibility for quality is devolved throughout the organisation.

Group Problem-solving and Improvement Teams

The use of the problem-solving team is a key element of the WCM approach to waste identification and eradication. The tools of quality and the creation of the team have been addressed already. One of the most important tools used by these teams is brainstorming.

This tool is widely known. It consists of getting your team together, identifying and defining the problem to be resolved and attempting to find the maximum number of possible ideas to address the question. A number of conditions are placed on the team and a number of techniques have been found to be useful over many years of use of the technique.

Conditions

- All are created equal
- All ideas are equal
- All comments to be positive.

Techniques

- Use a flip chart
- Pin or stick completed flip chart sheets around the walls of the room
- Split up the writing task between team members
- Have refreshments brought to the brainstorming room
- Apply pressure to participate to all members
- Hunt for suggestions.

The brainstorming session should be purposefully restricted in time, to increase the focus of all members. Once the ideas have been put on the flip chart they are stuck on the walls, in full view of the team. This allows members the opportunity to keep them in view as discussions develop. They act as triggers and idea creators as the day progresses.

When a sufficient quantity of ideas have been generated, the question of a rating system for the ideas arises. The quickest way to rate these ideas is by returning the team's attention to the lists and rating the suggestions as:

1. Definite — things that can be done *now*

2. Probable — things that can probably be done now, but require some further study

3. Possible — may have some chance of success, but not likely

4. Unlikely.

The reason for allowing unlikely suggestions to be made at the basic brainstorming session is that while any given idea may be unlikely in itself, it may well start a train of thought among the team, which in turn will result in a definite or a probable suggestion. Once the suggestions have been rated, the team proceed to create an action plan to assist implementation.

CONCLUSION

Henry Ford developed production to the point where very little if any of a worker's mental abilities were required to complete his job. A World Class operation needs the input and commitment of its people to achieve its goals of responding quickly and cheaply, with top quality, to its customer's needs. The role of employee involvement is central in devising and implementing effective World Class Manufacturing in the JIT and TQM areas. The logic of such involvement is irrefutable — two heads are inherently better than one — provided the effort has been made to train and develop people to understand the problems as well as the tools being used to resolve them.

Chapter 5

Business Process Re-Engineering

WCM IN THE OFFICE — BUSINESS PROCESS RE-ENGINEERING

In Chapter 4 we looked at appropriate structures for a WCM company. We looked at how information flows between departments. By applying the WCM principles of JIT, TQM and EI in the office environment, we can start to address the developmental opportunities available to us.

Business Process Re-Engineering refines the concepts of WCM for office applications. The activity starts with the basic processes of a company and looks critically to identify where value is being added. Some of these processes are universal:

* Order Entry

* Accounts Receivable

* Production or Service Supply

* Personnel

* Financial Accounting

* Accounts Payable.

Other processes are not quite as universal:

* R&D

* Marketing

* Sales

* Environmental Monitoring.

Unfortunately, only 25 per cent of time spent in offices has any value-adding input to it. This chapter will look at how we can reduce much of the typical waste found in offices.

In the production environment it is quite easy to see the product and process flows. Raw materials enter the plant, pass through various processes and leave the factory as finished goods. In an office environment it is often more difficult to see this type of flow. Our first difficulty is in coming to terms with the basic question: what is the product? If we look at letters produced, bills paid, accounts checked or orders entered as products we can start to visualise the processes within the administrative side of our companies.

The use of the JIT technique of process flows can help us in this area. When we complete the process flow we will undoubtedly see opportunities for improvement. As the people working the processes will be integrated into the analysis stage they will also be able to contribute to improving the system, thereby increasing their commitment to and ownership of the system they make work. Administrative staff can have difficulty in seeing their role in helping the company delight the customer. By integrating them into the overall company WCM programme they can appreciate their contribution to achieving the company's goals.

Flows

The process flow tool is useful when it comes to getting an objective view of what actually happens.

Figure 5.1 shows the physical flow of an order within this company. Many copies of the same information are made and distributed. The flow paths for orders received is presented as a thread diagram. Desk space was non-existent, with many staff having last seen their desks years ago. The company had a saying: "Give me a copy of that document before you shred it!"

Many of the people receiving copies of information didn't need the information and rarely, if ever, used it. By carrying out the process flow exercise office staff became aware of the ineffectiveness of their existing system and made a number of small but significant suggestions to management to simplify and improve the system. The simplified, improved system was implemented, allowing quicker response to customers, aiding the company in its efforts to improve customer service and loyalty. The office and administrative staff were integrated into the WCM effort from the earliest days of the company's WCM programme. This early and highly visible integration helped the company as the usual Them and Us barrier between shop floor and office staff was eroded due to the WCM initiative. People were able to talk across boundaries and to actively pursue improvements.

FIGURE 5.1: OFFICE FLOW

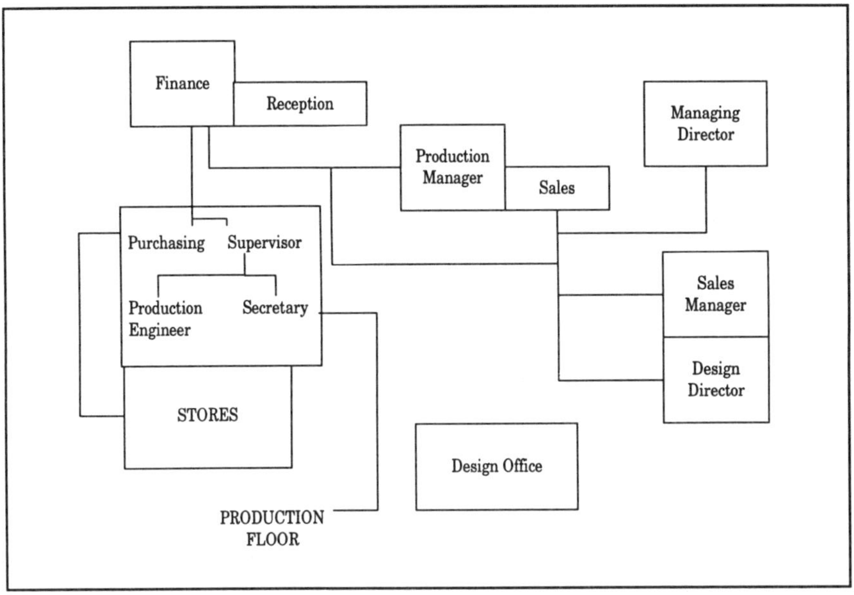

Waste Identification in the WCM Office

Our work in the manufacturing area has helped us to understand waste in a new way. We know that waste is any action that does not add value to a product, only cost. In the office environment some of these wastes go under different names:

- Moving it
- Sorting it
- Counting it
- Reconciling it
- Looking for it

- Checking it
- Reworking it
- Duplicating it
- Filing it
- etc.

When we carry out our physical flow analysis for a process we will record the steps involved. The theoretical and actual steps for a typical process such as entering a customer order is presented in Figure 5.2.

FIGURE 5.2: ORDER ENTRY — THEORETICAL AND ACTUAL

Theoretical	*Actual*
Receive Order	Receive Order — Reception
Enter into Sales Order Record	To Sales Office
Check Account Status	Wait
Place order on production	Enter into sales record
Plan into Production	To Finance
Produce	Wait
Ship	Clear account
Invoice	To Sales
	Wait
	To Production/Planning
	Wait
	To Production
	Produce
	To Shipping
	Documents to Administration
	Check papers with sales
	Wait
	Issue Invoice

The time spent waiting in each department is very much dependent on the work load of each individual and the way they do their jobs. They may process orders individually or wait until a number of orders have been received before logging them on the system. If they have a backlog, the time taken to action the document is completely controlled by this backlog. The situation is directly analogous to the departmentalised production area before the application of WCM principles.

When we examine the theoretical process steps and the actual process steps we see there is quite an amount of waste. If the order entry process is an important one for the company, shouldn't we try to make it work as smoothly as possible?

A simplified flow diagram for the traditional, theoretical process is presented in Figure 5.3. The offices are arranged in the traditional way with documents passing from Order Entry, to Finance to Planning etc., waiting their turn for attention at each office. This leads to delays in getting products manufactured and shipped to the customer. Another problem arises from the difficulty in accurately identifying where bottlenecks exist and where resources could be best applied to meet customer requirements.

FIGURE 5.3: TRADITIONAL OFFICE "FLOW" — ORDER ENTRY

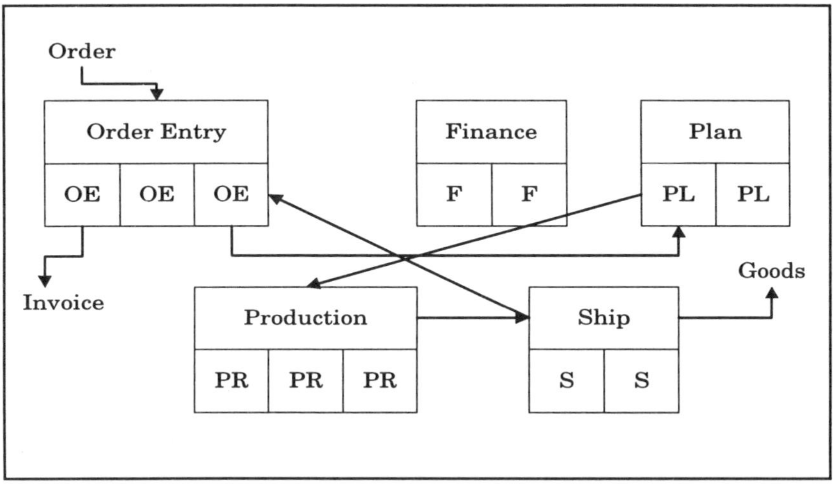

By grouping people together according to the process they support, as represented in Figure 5.4, significant improvements can be made. This concept is particularly applicable for processes that have been identified by the company as key processes, those which help to differentiate the company and its products in the market place — those processes which have a direct impact on the customer's buying decisions.

The application of the basic WCM principles leads to developing the actual processes to the point where it matches the theoretical best process. Now is the time to automate the process. Automation applied to a trimmed down process where the wastes have been eliminated will be both cheaper and easier to implement. Typical results from carrying out exercises as outlined above have resulted in companies moving from 2 to 3 weeks for order entry and acceptance to 5 to 10 minutes in the simplified WCM office.

FIGURE 5.4: WCM OFFICE FLOW — ORDER ENTRY

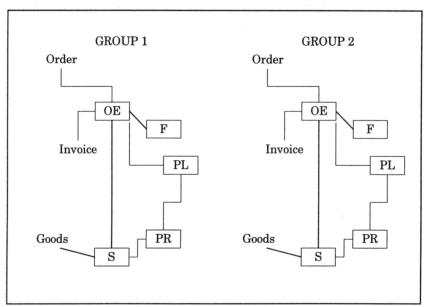

EFFECTIVE INFORMATION TECHNOLOGY

IBM launched the first PC on an unsuspecting world in 1975. Previously business had had to rely on large, cumbersome, expensive and temperamental IT department computers. Over the past 20 years the power, performance and general ability of computer systems has developed at a rate that was unthinkable back in those early days of Personal Computing.

Our typical applications have moved away from being DOS-based, capable of running from a single 360 KByte disk to being Windows-based, requiring up to 20 floppy disks for distribution and loading on your machine. A similar situation exists for the Macintosh and UNIX environments — system specifications and capabilities have developed exponentially over the last 15 years.

So what benefits have accrued to business from this development of raw computing power? Systems and applications have become easier to work, applications operate more reliably with cross-transfer of information now easily achieved from one programme to another. More work can be done by fewer people — in theory at least.

Before we look at how IT is used by World Class companies, let's take a brief look at the operating systems and applications currently in use.

Operating Systems

The hardware is only one part of a computer system. The operating system is the controlling brains that interfaces between the hardware of the system and the applications software being run by the system's user. The most common systems would be:

1. *Microsoft Windows*: Microsoft Windows is a Graphical User Interface (GUI) which allows users to run applications by selecting pictures or icons and effectively buffers the user from the internal mysteries of their computers. The Windows environment was often confused as an operating system, but it has effectively captured the market for normal office and home users around the world. In much the same way as the modern car, Windows insulates the user from the working components of the machine. The GUI allows simple access to quite advanced features of the computer which were previously only accessible by data processing professionals or enthusiasts. A key feature of the Windows environment, also like the modern car, is that you really don't need to understand these features on a daily basis, you just use it and it works effectively. The system takes care of them for you, being delivered with "Plug and Play" easy installation and system configuration software. Windows has effectively brought the power of computers to the masses.

2. *MS-DOS*: Originally crafted by Microsoft as an operating system for the IBM PC and sold as PC DOS. The system is a single-user variety, and is single tasking i.e. it does one job at time. The system was originally designed to handle a maximum of 640K memory. The exponential rise in computer per-

formance has stretched DOS to the limit. The modern day computer user has little need to enter the DOS level of their machine, the Windows environment having greatly simplified the majority of tasks that DOS was needed for.

3. *OS/2*: IBM's own operating system. This was designed as a multi-tasking, single-user system. However, its widespread acceptance in industry has been held up, probably because of the introduction of the Windows environment by Microsoft.

4. *System 7*: Apple's latest operating system for the Macintosh environment. The Apple systems have been designed with a high degree of attention given to the user interface. Where DOS was keyboard based, the Macintosh interface has been graphically based from the outset, with the user using a mouse to select tasks etc. Apple were the first large computer company to gain wide market acceptance for their graphical interface. The system operates as a multi-tasking, single user system, and was originally based on the Motorola 68000 series of chips. Note: the multi-tasking feature on single user PCs has been available for some years, having been available commercially on the Sinclair QL.

5. *UNIX*: UNIX is probably the most widely known, widely awaited system providing multi-user, multi-tasking features. The system is considered to be extremely powerful and capable but has been "expected" to take its place of prominence in the computer world for the last 20 years. Many variants of the system exist, with a good level of understanding required to use the system effectively.

What is universally constant for operating systems into the future is that as computers become more powerful, the operating systems themselves will be required to handle ever more complex systems. The borderline between operating systems and such systems as Windows and the new Windows NT will become more blurred. Windows NT appears to offer fixes for the inherent problems associated with DOS. An important statistic places the operating systems in perspective:

80 million people use Windows
11 million use Apple Macintosh (System 7)
5 million use OS/2
2 million use UNIX

Applications

As machine power has increased so also has the power of the applications software. The software/machine combination allows so called "Power Users" to do many things they would have found impossible ten years ago and very difficult five years ago. One point often missed by management, however, is that the vast majority of computer users are *not* Power Users. Most people do not need the computing power of a Pentium machine running at 120 MHz to run a word processing package or a basic spread sheet application.

On the other hand, certain industries have been completely revolutionised by the introduction of new technology. The publishing industry, for example, has managed to remove three or four full steps from the production process through the use of advanced processing power. The benefits achieved in this case result from examining the totality of the process rather than suboptimisation of individual tasks.

The true benefits of the new computer technology will become clear when the systems are examined in their totality — when WCM principles are applied to the system and system development. Too often over the past 20 years, technical solutions have been applied just because they were available rather than because they contributed to simplification of the overall process. The major thrust in computer development in the next decade is believed to be in the area of portable computing. As computers become smaller while retaining their power, they become more useful. When the portable computers are allied to the developments in communications the next major step in system ability will take place.

As computers themselves have become more powerful, so too have the applications running on them. It is common today to find word processing packages in normal offices that would have represented the pinnacle of printing house technology five years ago.

Until the widespread availability of CD-ROM drives it was usual to have to load over 20 floppy discs into your computer to install a major new software package. The CD-ROM now allows the same software to be distributed on a single disc. In the same way that few users use the full capability of their computers, even fewer use the full capabilities of their software. Most users have a quite limited requirement, often satisfied by quite simple elements of their software. When Microsoft were working on an upgrade to their "Word" word processing programme they carried out a lot of customer survey work, to find out just what "new" features their customers wanted. They created quite a long list of new, desired features, 95% of which were already supplied on the current release of the software. The introduction of the Internet is offering a potential solution to this problem of cost and size of software packages.

Telecommunications

Telecommunications appears to offer the next major leap forward for computers and computing. Systems such as the Internet, an on-line information service, appear to be growing at an exponential rate. Users can now gain access to information, worldwide, at a fraction of the cost of obtaining it in a more traditional way. Communications developments have enabled this move forward. Telephone networks have largely been converted to digital exchanges, with the advent of Integrated Services Digital Network (ISDN) systems expanding daily. Such ISDN systems allow the reliable transmission of large amounts of data (min. 128K Bits/second). This is approximately 9 times the capacity of a normal analogue telephone system. The ISDN systems make the transmission of video action possible over the telephone!

The use of the Internet and the World Wide Web (WWW) is currently sweeping through business. The value of the Internet economy has grown from zero, when the system consisted largely of academic community links, to today where it is valued at $10 billion. The rate of growth of the Internet is phenomenal with 12,000 users being added to the Web community every day. Most of today's computers are delivered "Web ready", with modems installed and sound cards fitted. The system allows for access to information on topics from West Highland Terriers to astrophysics and the latest advances in science, business and technology.

Most of the world's leading companies have a presence on the Web, where they provide corporate information, product updates and often, technical support. Companies such as NetScape have built businesses because of the Web. Software companies in particular have taken to the on-line world of the Web with enthusiasm. They can cost-effectively market, sell, distribute and support their products on this new medium. This facility to distribute software and upgrades over the telephone network is leading to a new type of software sale. "Applets" are reaching the market: software manufacturers create a piece of core software (in the case of a word processing package, this would be the central programme to allow basic typing functions, corrections and formatting); this core programme can then be upgraded by the customer with additional features such as mail merge, advanced fonts etc. These upgrades are known as applets (little applications) and they can be down-loaded over the telephone line via the Internet, allowing the customer to match the software being bought to the needs and pocket of the customer.

A similar approach is being developed by the hardware manufacturers. A number of companies are developing Netputers, machines designed specifically for working in conjunction with the Web. These netputers are quite simple machines and can therefore be made and sold relatively cheaply, at about 50% of the price of a full computer. They would be similar in concept to the terminals we would remember from the days before the PC revolution. Their success will probably depend on the ability of Web service providers to support their distributed computing needs.

The Impact of IT on World Class Companies

A key factor in the use of information technology is that despite significant developments in the applications field, there have been limited improvements in the overall efficiency of operations and productivity. Studies in the US, for example, have shown that despite a 50 per cent increase in investment in IT equipment and software in the period 1975 to 1994, office productivity has increased only marginally, by approximately 5 per cent. This lack of payback on investments made appears to result from a traditional approach to process improvement rather than the World Class approach.

Further analysis has shown that in a typical office only 5 to 10 per cent of the time a document spends in the office is value-adding, that is, actually having work done on it that a client or customer would be willing to pay for. The rest of the time it is being sorted, distributed, checked, etc. This is analogous to the situation we have seen on the traditional production floor. Papers, invoices, production plans etc. spend time waiting in different departments, being moved, shuffled, checked and double checked as they pass through the office system. An effective use of information technology, however, can now be used to address these wastes in the office systems.

In the traditional developmental approach an analyst would look at the existing manual system and attempt to automate it. Seldom was a root and branch analysis carried out of *why* the job was being done in the first place or *who* was also involved in the job outside the specific department being analysed. In effect, the waste inherent in a system was automated. This meant that the waste became endemic and deep-seated in the organisation's systems and culture. This helps to explain why despite vast investments in information technology, overall productivity has only improved by approximately 5 per cent.

Information Acquisition
World Class companies respond quickly and accurately to customer needs. In normal business the customer satisfaction process actually starts with the initial sales contact and finishes when the invoice has been paid. To meet customer needs quickly salespeople on the road are now frequently equipped with some form of mobile communications and computer link. In the retail trades particularly, portable computers with communications such as fax modems are used to relay customer information and orders directly to suppliers' systems. This simple link has managed to remove many steps, actions and intermediaries from the traditional order processing area.

The ultimate level for World Class companies is attained when direct links are made between a customer's point-of-sale equipment and the supplier's system. When we visit our local super-markets we see part of this system. By using bar codes and readers the sales data are read into the shops computer as the sale is made. Stock levels are immediately depleted on the retailer's stock system and orders are placed with suppliers as necessary. Staff no longer have to price goods individually, or key them into a cash register, or count individual stock items on the sales floor, or manually deplete stock levels in a stock control system, or prepare orders for suppliers. Thus the modern supermarket operates in a World Class way. In fact, in the 1950s, many of the basic concepts which have since been collected under the Just-in-Time banner were originally created in mass retailing operations in America. As technology such as bar coding, integrated stock control systems, intelligent cash registers and computer telecommunications have developed, so too have their widespread application in business.

Information Transfer

Gathering the information is only one part of the process. To be useful, the information has to be transmitted and received in a timely manner. It is probably in areas of communications that the most impact will be felt in the next 10 years, and World Class companies are at the forefront of this revolution. By linking into the point of sale and customer stock control systems, suppliers can greatly reduce the lead time associated with the replenishment of sales stocks. The lead time traditionally associated with raising an order involved creating the order, posting it, receiving it and entering it into the supplier's system. Telecommunications have immediately allowed the removal of two days from this process through the use of fax machines at the basic level, to the use of Electronic Data Interchange at a more advanced level.

Developments in telecommunications are enabling these delays to be removed from the operations of World Class companies. As each delay is removed then so is the amount of stock required to buffer the system. As the stock levels are reduced so also is the amount of cash required to finance the business. Time is money.

The introduction of ISDN services for fixed telephones and the GSM (*General Système Mobilier*) for portable telephones offers the possibility of significant developments for both computers and telephones. As the GSM system becomes more widespread, it will be simpler and more effective to use portable computers "in the field". Additional features such as cellular, satellite and radio communications will be incorporated into portable computers allowing access to information systems without recourse to an office.

So how can this technology be utilised by World Class companies? As sales, engineering and support staff will be able to communicate freely and effectively, lead times to market will be reduced. Current technologies allow a form of computer-based teleconferencing, where people located widely apart can look at the same drawing or document and make notes, suggestions and comments directly onto the drawing that are immediately replicated on all the machines at the same time. This can be very useful when it is difficult to get all members of a team together at the same place and time. As the telecommunications systems now allow not only voice and pictures to be transmitted but also technical drawings to be edited interactively, the potential for further reductions in lead times to market for complicated products is enhanced. Wastes and delays can be further removed from the product introduction process allowing World Class companies to react even quicker to customer requirements.

Information Storage and Retrieval
Business demands the ability to supply not only the goods or services that a customer wants but also the supporting paperwork. The introduction of computer databases have allowed companies to keep client account information available to customer support staff. However, in certain industries, details of transactions are only a small part of the process. In the banking and insurance industries, for example, hard copies of medical certificates, marriage certificates, land registry documents and a myriad of other documents need to be stored and available for access at short notice.

The introduction of Imaging Technology has enabled many such institutions to store and retrieve such information as easily as the more traditional customer account details. Institutions like the ICS building society have invested heavily in the technology and this has allowed the company expand inexpensively. Staff are able to perform to a level unprecedented in a more traditionally-based institution. Documents are scanned electronically and stored in a file. A staff member is then able to retrieve an image of the document without having to go looking for a physical document among thousands of files. Imaging technology is supported by a number of separate but interlinked technologies such as scanning, data compression and optical data storage and retrieval.

CONCLUSION

The power of Personal Computers have only been available to the majority of industries since the mid-1970s. In this short period of time computing power has become an essential element of our daily lives. From the supermarket check-out to the production floor, computers are highly integrated into our society. One of the problems associated with computers and computing is the speed of change in the industry. Once a new computer has been introduced it is usually only a matter of months before it becomes obsolete. It's only three years since a 486-based computer was regarded as a highly desirable machine. Now there are processors on the drawing board, giving levels of computing performance only dreamed about five years ago. These chips hold great possibilities for the future of computing.

The future for information technology is particularly difficult to predict. What has been outlined above already exists and is available at least to the enthusiast and the power user. The future will see an increasing use of these tools, as they become easier to use and their prices continue to drop. One thing is certain: World Class companies will continue to be at or near the leading edge of these technological advancements, looking to see what new tools can be used to help them identify and remove further wastes from their operations.

Chapter 6

World Class Design

The design of products is a fundamental area for many businesses. The basic design and specifying of a product can account for over 70 per cent of the total costs associated with a product. Seemingly small variations in specification or production processing requirements made at this stage of the product life can have a disproportionate impact on the life cycle cost of the product. The graph shown in Figure 6.1 represents this fact. The Committed Spending on a new product rises rapidly during the design and prototype stages. At this time the designers have decided on the basic design, the parts to be incorporated into it which effectively defines the majority of total product costs in many cases. The Actual Spending line shows at what stages the actual cash is spent in creating a new product. Much more money is spent on the product at the time of production as compared to the idea and prototype phases.

FIGURE 6.1: COST OF DESIGN

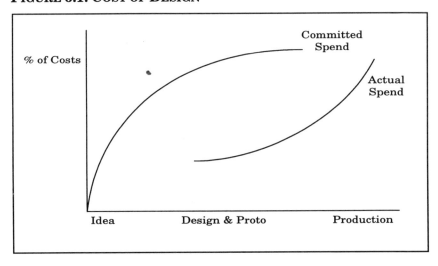

For this reason designers, marketing, financial, purchasing and production staff need to communicate well during the initial design stages of a product to ensure unnecessary costs and features are not incorporated at the early stage of the product's life.

PRODUCTS AND SERVICES DESIGN

The past 10 years have been very difficult for Western manufacturers. Japanese companies have taken the market by storm, taking lead positions in many market sectors, such as cars, VCRs, stereo equipment and so on.

Product life cycles in the car industry which were measured in 10 year units have been cut to the point where the total life cycle is now down to three years for a completely new model. The Japanese were able to design and introduce a totally new model every three years whereas the better Western companies took seven years. As the life cycle was reduced, the competition increased, as new features and technologies could be offered for sale as the new models were introduced. Sales of Western manufacturers fell as those of the Japanese soared.

In the personal entertainment goods arena, Sony is a household name. The ubiquitous Walkman has a life cycle of three months in Japan. For a time it appeared as if the Japanese would be the only manufacturers of the world's goods! The European manufacturers have responded in the past 5 years, competing on design, performance and now quality against the best of the Japanese. The Europeans have taken the concepts of WCM to heart and are now able to compete equally with the best. Companies such as Philips have introduced company-wide quality and process improvement programmes. They have intensified their efforts at team building and employee involvement. Many of the previously mentioned core tools and techniques of WCM are applicable in the products and services area; however, there are some tools specifically designed for this function.

The ability of Far Eastern competitors to shorten product life cycles and simultaneously grow market share puts Western companies under extreme pressure. The design processes used in Japan and the West have been studied and the following categories have been identified as being critical areas of difference between the two cultures.

Leadership

Most large-scale design projects are carried out using a form of matrix management. A project team leader is appointed who forms a team with relevant expertise, usually from the marketing, finance, production and engineering departments.

In Western companies the project team leader operates as a co-ordinator. They have to try and persuade departmental heads to allocate resources, to free up people to ensure projects proceed to schedule. The project team members are "on the team", but still wholeheartedly committed to their department and to furthering its own goals. Their career paths are closely linked with the Department. Success or failure of the project team is not that critical to them.

The Japanese have a different concept of Project Leader, the *shusa*, or boss. Toyota were the pioneers of this concept, as of so many others.

The *shusa* is the team leader. They are responsible for getting a product or service designed and produced. They will take the project from the start to the finish and have wide-ranging powers within the company to ensure its success. The *shusa* will operate the project team under a Matrix Management arrangement. Staff will be made available to the *shusa* for the duration of the project. Their performance in the team will be critical to their career development as the company places great emphasis on the area of new product development. They will retain links with their departments but with a view to getting the best result for the project team and the company rather than for their own department.

The role of the *shusa* is very highly regarded in Japan. Many of the leading manufacturers attribute specific products to these individuals. They have effectively taken on the persona of Super Craftsmen, working through their teams to craft products using the full resources available. They are the driving force behind the mass craftsmanship of the Japanese.

Teamwork

The concept of *shusa* appears to explicitly contradict all we know of close teamwork. We have an understanding of close-knit teams where each person's individuality is secondary to the team. In the *shusa* system the leader leads the team in a well-orchestrated and co-ordinated way. Team members are respected for their individuality, as contributors to achieving the goals of the team. The major Japanese companies have developed the team system to the stage where heavyweight teams, *shura*, are used on particularly important projects. The effectiveness of a team is measurable by its results, as shown in Figure 6.2.

FIGURE 6.2: PROJECT TEAM MANNING LEVELS

	No. of Engineers Employed on Project
American / European	900
Japanese — *shusa*	500
Japanese — *shura*	330
European	1,400

Source: Womack, et al., *The Machine that Changed the World*

Communication

Western companies communicate on a superficial level, at least at the start of projects. Discussions take place between the relevant departments of Marketing, Production, Finance, Engineering, Planning etc. but rarely do the hard questions get answered. Projects will be started and developed without fully detailed specifications and parameters being agreed until the product is nearly ready for market launch. In the Western company the number of people involved at the beginning of a project is small with often large numbers involved as the project reaches a conclusion.

The Japanese *shusa* system operates differently. The hard decisions are taken early on in the project. Basic parameters and specifications are defined and agreed. Marketing and design staff agree the concept, the features and the performance characteristics of the product. Finance, planning and purchasing know the implications for them of the new product and can plan and prepare for it. The number of people involved at the early stages of the project is high. As the hard decisions are being made, the numbers fall off as the project reaches a conclusion.

This breakdown of project staffing levels is represented in Figure 6.3. The Japanese have lots of team members present and involved at the early stages of a project, to get answers to the hard questions. The Westerners have few team members present at the start, building up to maximum levels as the product reaches introduction point.

FIGURE 6.3: PROJECT STAFFING — EAST VS. WEST

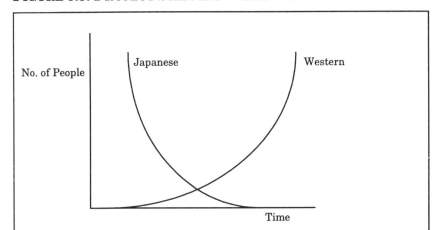

The *shusa*'s job is to force the group to reach a consensus. The difficult decisions are faced and made at an early stage. People can then concentrate on achieving a specifically defined goal rather than working on a number of options that may or may not be required.

Simultaneous Development

The traditional design process results in sequential development, as represented in Figure 6.4. Designs are completed before packaging work starts, before production processes are designed, before materials are sourced etc. This results in a build-up of lead times as each downstream department has its own lead-time requirements to complete the specific stages of the new product introduction process. Any changes in design or specification can have a significant impact on the introduction date of the product. Often production and purchasing staff will hold off making purchasing decisions till the last minute because of fear of product design changes.

FIGURE 6.4: TRADITIONAL DESIGN PROCESS

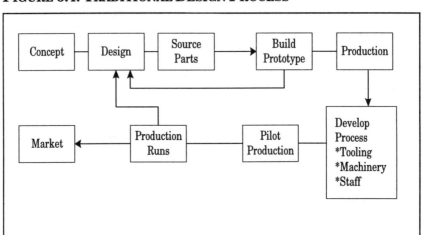

You don't have to do all the design jobs, one after another. Provided the communication levels are good it is often possible to do significant amounts of preparatory work in the downstream areas before a design is completed. Provided production is aware of the outline parameters required of it by new models, it can prepare itself for the new arrival. Provided plastics moulders or metal fabricators are aware of the outline requirements of a product they can prepare basic tooling ready for the new product. Simultaneous development can work where communication levels are good. Figure 6.5 demonstrates this concept.

By ensuring that communication levels are good, individual team members can make decisions early and well to short circuit the overall product introduction cycle. Design and Marketing need to work closely together so that designers truly understand the needs and desires of the end users. As this closeness develops a company's products are more closely attuned to the requirements of the market place and so are more likely to be bought.

FIGURE 6.5: WCM DESIGN PROCESS

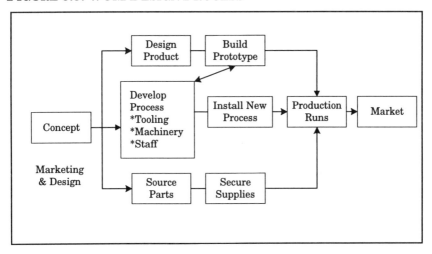

Design for Manufacture

A well-designed product with lots of marketable features that cannot be manufactured effectively is a poorly designed product. A company designs products to satisfy or exceed customer needs and requirements. But the company also has needs and requirements, it needs to satisfy its stakeholders, it needs to make a profit, it needs to continue in business and develop. If the products it designs cannot be manufactured profitably, then it will fail.

Modularity

The design for manufacture concept works to help companies design products that can be manufactured effectively. Back in the 1930s and 1940s, General Motors came to understand this concept. Many of their models were designed using main modules that were then packaged in different outer skins. We see the same happening today with Saab offering GM engines in a number of models. By using well-proven designs, by using tried and tested modules, a company can significantly reduce a product's development time as well as reducing the complexity within the production area when it comes to manufacturing the product.

Standardisation

If a company can standardise a number of modules or sub-assemblies across its product range significant savings can be made at the production level. Designers want to design; it's their basic function, but their creative talents are best utilised when they design using standardised components or parts. The range of parts available today means that any number of designers can design a product to meet a given specification without any two of them using the same components from the same source. It is fundamentally important to put some basic controls on the type and variety of components available for choice by the design team.

Part Count

The number of parts required to make up a product is a very important consideration when it comes to manufacture. Figure 6.6 shows how a bolting operation can be significantly simplified by reducing the number of parts required. In this example, the number of parts required has been reduced from four to two. The electronics industry has been a front runner in this quest for part count reduction.

FIGURE 6.6: REDUCING PART COUNT

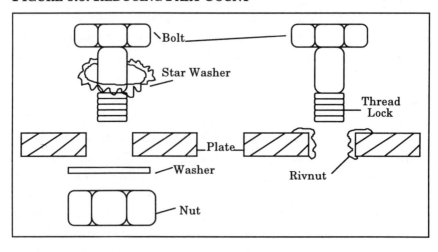

By reducing the number of parts you:

- Reduce inventory
- Reduce handling
- Reduce suppliers
- Improve quality.

The fewer the number of parts the better the quality opportunity.

Design for Assembly

The previous example of reducing part count has an equally important application in the area of design for assembly. By reducing the number of parts the labour content of the assembly operation can be reduced.

By designing products with a view to how they will be assembled in production, designers can have a significant impact on manufacturability, ensuring products can be assembled from the bottom up, avoiding the necessity to turn them over for access. Minimising the number of parts and of bolted joints, replacing bolted assemblies with snap fits, improving alignment systems are all useful techniques to improve production performance.

Applying the principles of POKA YOKE, a Japanese concept of fool proofing, at the assembly stage can save large amounts of lost time and waste further along the assembly line. *Design products so there is only one, obvious way to assemble them.*

Design for Testing

The test function is quite often a bottleneck in a manufacturing operation. Parts or covers need to be removed, test points are difficult to get at or new, unproven sub-assemblies or components have been incorporated in a design requiring extensive testing of products in an effort to ensure product quality is not compromised. Designers should design to allow for quick and effective testing.

Design for Process

Simplicity has been a constant thread running through the tools and techniques of WCM. When designing a new product consideration should be given to how it will be manufactured, to the physical processes required to get it made. Unless specific technological requirements demand it, it is best to stick to tried and proven processes. Proven equipment can manufacture with a degree of consistency not always obtainable from new-age technologies. Frequently, the winning companies let a technology pass the experimental and first rush stages before incorporating it as an integral element of their core production processes.

Product Evaluation

It is important to know what is happening in your market. When a new product is launched by competitors it is very important that your staff get a good understanding of it as soon as possible. The story is told that the first three samples of each new Toyota model are bought by Nissan, Mazda and Mitsubishi. In Europe, each of the main truck manufacturers issue their sales staff with competitive analyses of their competitors' products.

Close scrutiny of competitors' products can pay dividends for your company. You may see a new idea that, when modified or developed by your own staff, may lead to significant improvements and cost savings. The product evaluation process should be organised and professionally carried out. Ideally, your own customers' requirements analysis should be used when examining competitive products, to see where the competition rests in relation to your own features and characteristics:

FIGURE 6.7: COMPETITIVE ANALYSIS

Feature	Competitor	Your Product
Price	20,000	23,000
Power	150 HP	165 HP
Capacity	10T	10.2T
Annual Sales	250	150

By critically examining the features of competitive products you can ascertain how well your own product meets the customers' requirements. When sales volumes are factored into the analysis you can come to an understanding of the sensitivity of the market to different factors and features. This is represented in Figure 6.8, where the annual sales for your product and your competitor's are multiplied by the retail price. This multiplication gives a figure for the Relative Market Worth achieved by both your own and your competitor's products. This is a simplified version; obviously very few market segments have only two competitive products. A full review of your competitors using this analysis tool can often be very useful.

FIGURE 6.8: RELATIVE MARKET WORTH

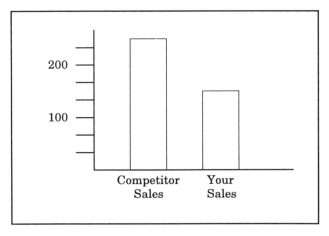

This simple analysis would seem to indicate that the customer places more emphasis on price than on additional features and performance. This would be a starting point for a re-examination of your interpretation of your customers' buying criteria, and possibly of your positioning in the market.

"Best of Breed" Benchmarking

Benchmarking is the name given to a process of making comparisons between your own company and its processes against those of your competitors and other companies not operating in your market, but which are acclaimed for their superiority in particular processes.

In the early days of benchmarking, companies focused on their own performance relative to direct competitors. In the recent past companies have looked further afield in an effort to compare particular processes within their organisation with the "best of breed". By looking to compare your processes with the very best you are coming close to having an objective view of your own performance.

The earliest form of benchmarking was exercised by the financial people, through the application of financial ratios, to financial results published by companies. Industrial engineering also created standards that could be applied to determine how well a company was performing particular tasks.

The main difficulty in applying a benchmarking approach to a company is in determining what benchmarks are appropriate. The basic requirements identified during the buying criteria and selling criteria analysis can be used effectively to prioritise areas for attention and also to help identify appropriate benchmarks. If the buying criteria indicate that price is the key customer requirement then total product cost should be addressed as forming the basis for a suitable benchmark for the company. The market leader on price, given that this is the key customer requirement, should be selected for analysis.

If the key buying criterion is speed of response then the organisation selected for comparison may well shift outside your market and possibly use a company such as the AA, Rank Xerox or McDonalds as suitable benchmark companies.

To carry out an effective benchmarking exercise requires ruthless objectivity. It is essential that a totally objective view is taken of your own process to ensure the conclusions drawn from the exercise are valid. Armed with this objectivity we can now carry out the benchmarking exercise:

Stage 1: Plan
- Select the processes within your organisation to be benchmarked.
- Select the benchmarking criteria/measures

- Select the comparison company

Stage 2: Do
- Do a comparison of your company's processes against the chosen comparison company or companies

Stage 3: Review
- Review the comparison studies
- Identify improvement possibilities
- Plan remedial actions

Stage 4: Implement
- Put the plans for improvement into place

Stage 5: Feedback
- Analyse the improved processes
- Verify that the chosen processes have improved and that they were in fact the key processes for your company.

The benchmarking approach can be very useful as an overview approach for a company. Certain difficulties occur at the micro level as you try to understand the detailed complexity of another company, often operating outside your core business, without a full understanding of the internal interactions of that company's structures and objectives. At the macro process level, it can provide useful guides and independent measures of performance for managers to evaluate their performance against.

Chapter 7

Financial Analysis for WCM

The term World Class Manufacturing makes most people think primarily in terms of manufacturing, that WCM is only applicable on the production floor and only focuses on operational matters. In part this is true; WCM does concentrate on operational issues, but not only on the production floor. It looks at a company in a holistic manner, right across the organisation. WCM concentrates on operational issues because they have a direct impact on a company's bottom line. By removing internal wastes, by reducing unnecessary costs, a company can significantly improve its profit. Wastes cost profit so it would be fair to say that WCM concentrates on profit. Because of this focus on profit, it is important that anyone starting to work with WCM understand at least some basics about finances, accounting measures and how to interpret information your business is trying to give you.

MEASURES FOR WORLD CLASS MANUFACTURING

The old adage says: "What we measure, improves!" When implementing a WCM programme a number of measures of improvements made are very useful. Once again, however, it is important for you to interpret measures correctly. While some measures are practically universal, others have only limited applicability. Choose measures which have an impact on your business, on your ability to service customers, rather than just choosing measures because they are "standard".

One of the difficulties typically encountered when applying traditional measures to monitor a WCM programme at its early stages of implementation is that things may appear to be getting worse rather than better! At the early stages of a JIT programme, for example, stocks may well rise, as a two bin system is put in place. It is only when the system has settled down and confidence levels have developed in the ability of suppliers to meet requirements that real stock levels will begin to drop. What *will* have improved will be the levels of output, the efficiency of production and, most likely, the quality and customer service levels. In a recent WCM implementation exercise, which in fact was going very well, customer service level "failed to improve". This was at variance with the reality on the ground. For example, customers were no longer constantly on the telephone looking for stock — even though shipments had increased by the company! The "problem" was that customer service levels were monitored on a monthly basis while customers wanted deliveries as they needed them. It took a number of months to balance these two factors, until measured customer service levels matched the reality on the ground, where customers were being effectively serviced.

Similar situations arise when starting on TQM and EI programmes. The costs associated with identifying the root cause of problems are borne up front, while the benefits are not seen for a time. So how can we see if our efforts to implement World Class Manufacturing are worthwhile?

At the outset we identified that WCM was about removing wastes, which in turn increased profitability, so possibly the first measures should focus on *profitability* and *waste*.

A measure of the profitability per employee will tell you quickly whether you are improving the efficiency of your operation. As profits are usually calculated at the end of accounting periods, a more immediate measure of improvement is the turnover per employee. This measure is relatively easy to calculate and is generally immediately available.

Waste is more difficult to measure, in some ways. At the most basic level it is possible to count and cost the amount of waste or reject products manufactured in a given period. We know, however, that this is just the most obvious measure of waste, and efforts should be made to identify and quantify the other wastes we know to exist within our operations. Armed with this full cost of waste, management is in a good position to decide on investment and development decisions.

In our simplified definition of World Class Manufacturing we said WCM was about making products *Quicker, Better and Cheaper . . . Together*, so we should look for measures that will tell us if we are achieving these objectives.

Quicker

In a World Class company we want to manufacture our products quicker to allow us to satisfy our customers' needs as quickly as possible. The first and most basic measure we can make to monitor our speed of response is to measure our lead time. In the case of measuring customer satisfaction we will define lead time as the time between a customer placing an order and receiving the goods, correctly documented and packed. As we have seen there are other measures of lead time that World Class companies use, such as the time to receive goods from suppliers, to get bills paid, etc., but we will focus on the above definition for the moment. This measure of lead time encompasses a number of areas within the company:

- Sales staff

- Order entry

- Production

  — Purchasing

  — Stores

  — Manufacturing

- Shipping

- Administration.

Merely defining the inter-related departments within a company begins to give some indication as to possible delays in the customer satisfaction process. And by measuring the lead time on this essential process we can start to look for ways to reduce it and improve our levels of customer satisfaction.

Some companies not only manufacture products but also design new ones. In these companies an often critical lead time relates to the product introduction cycle. Once again, by listing the relevant departments and closely examining the process, management will start to identify value-adding steps, as well as those which only add cost and time delays to the process.

World Class companies can typically achieve significant reductions in lead times, e.g. from 4 to 6 weeks to 1 day. Measures of on-time deliveries can be useful in helping a company measure and subsequently act to improve customer service.

Another simple measure used by World Class companies is process flow, or the distance parts, products and people must travel to get through the production process. In a recent example, an operator made 1,000 items a day and had to move ten feet each time to the next operation. His round trip was 20 feet, which doesn't sound like a lot, but in a day he walked 3.8 miles! The operator had effectively been spending 14 per cent of his working day *walking*. This simple measure is very useful to a World Class company as it develops its processes and its systems.

Better

World Class companies make products better. The traditional measures of quality — such as reject quantities, customer complaints etc. — are still valid in a World Class company, but they need to be used actively and aggressively to improve systems, processes and products. Rates of improvement are very important in these measures and can be charted by the people responsible for them. It is often quite a simple matter to quantify the monetary value of rejects and this should be done to help people realise what rejects are costing the company. For example, one client was experiencing less than satisfactory profitability. When he informed his staff that the value of rejects and breakages within the manufacturing operation was equivalent to the *total* profit made by the company, the number of rejects and breakages dropped dramatically. When people are made aware of the value of the waste they can help to eradicate it.

Cheaper

World Class companies are able to produce products more cheaply than their competitors. Costs can build up in many ways, from basic design choices to manufacturing and distribution decisions. This book has attempted to identify wastes and guide managers and staff to enable them to remove them. A number of specific measures in relation to monitoring the "cheaper" part of World Class Manufacturing are now presented.

Product Cost

World Class companies monitor their basic product costs. They endeavour to reduce these costs at the earliest stages of product design through the use of a number of design tools and techniques such as Value Engineering, product standardisation etc.

Stocking Costs

World Class companies look for costs all along the value chain. When they look for stocking costs they start with their stores areas. As their ability and systems develop, many of them move to the point where they no longer have formal stores areas but instead keep parts at the point of manufacture. These points of use are then re-supplied by the part supplier directly.

Until they reach this point, however, a useful indicator of materials control improvements is the measure of stock turns. As stocks can exist in a number of different forms in a company, there are a number of measures for stock turns. The most basic figure for stock turns occurs at a global level. It relates the stock levels to the volume of sales.

Stock turnover can be defined as follows:

Opening Stock – Closing Stock = Subtotal

Average Stock = Subtotal ÷ 2

Stock Turnover = Cost of Sales ÷ Average Stock

The figure is presented as the number of times stocks are replaced during a year. In a World Class Manufacturing company the initial target for stock turns should be at least 12, with the best manufacturing companies achieving stock turns of over 100 per year.

This measure of stock turns occurs at a global level. In the World Class company we look at the large picture and then focus in detail on the specifics. A key measure of stock turns and stocking costs should and must focus on the levels of Work-in-Progress (WIP) within a manufacturing operation. Here a simple measure of the cost is extremely useful in helping management identify and prioritise areas for improvement. A measure of WIP turnover can be used:

WIP turnover = (Avg. WIP X 365) ÷ Production Cost of Sales

Once again the World Class Company will monitor these figures and look for trends to check how they are improving their operations. One of the simplest and most useful tools in this area is to produce a layout drawing of the operation and mark on it the values of stock being held in each area. This simple method allows managers to see where their attentions are best focused.

World Class Companies are obviously highly efficient organisations. A very simple measure of this efficiency is the amount of sales revenue that each employee can produce. This is usually referred to as turnover per employee. It can be calculated using the following formula:

Turnover per Employee = Sales Revenue ÷ Number of Employees

This figure is easily obtained and has a great significance for a company. In "normal" industrial development programmes, companies usually achieve 2 to 7 per cent annual improvements in this figure. World Class Companies often achieve over 20 per cent rates of improvement. Because the measure is so easily calculated, it can be determined on a monthly or weekly basis, giving management a readily available indicator of how well their operations are performing and what the trends are like. The measure can be very useful in determining if recent appointments have had a beneficial affect on overall throughputs and if new systems are performing well. Many manufacturing managers apply the measure just to the manufacturing areas, where they have control and responsibility. It is an equally valid measure at CEO or General Manager level to monitor the effectiveness of the overall organisation.

Together

The World Class principle of Employee Involvement is fundamental to the success of a company's efforts. Measures of the morale of a company's staff are often quite difficult to obtain and monitor, especially for a small company. But there are a number of specific measures that can indicate how workers see the company and their jobs.

We have already addressed a number of these from the company's point of view — "Is productivity improving?" "What about quality and lead time and rejects?" and so on — but from an employee's point of view the question might be, "Is the company a good place to work?" Simple measures to gauge the morale of the employees would be staff turnover rates, absenteeism rates, and hours lost.

For example, a client was having severe staffing difficulties, with high turnover rates, high absenteeism and difficulty in getting people to work there. An analysis of lost hours due to absenteeism made very depressing reading. This meant that management had difficulty in meeting production requirements as staff were not available to work. Four months after starting to implement World Class principles, the absenteeism report and staff turnover rates had improved dramatically and the company actually had people applying for jobs in the factory.

These are simple measures and not explicitly related to staff morale, but they are legitimate measures of how people perceive their workplace and their jobs.

This section on World Class measures is not intended to re-place a number of the traditional financial measures for monitor-ing company performance such as:

- Return On Net Assets

- Liquidity Ratios

- Profit Margins

- Debtor/Creditor Days, etc.

and the many more highly useful and often simple to use meas-ures that have been developed by the accounting profession over the years. If these measures mean little or nothing to you there are a number of worthwhile books available specifically dedicated to accounting for SMEs. These basic measures should be studied and understood to allow you to effectively use much of the infor-mation that your business can tell you.

These are but a few of the many measures commonly used in a World Class Manufacturing environment. The next section will introduce a concept known as Activity Based Costing, where ac-countancy systems align themselves with World Class Manufac-turing.

ACTIVITY-BASED COSTING

The use of the term World Class Manufacturing tends to affect people in different ways. Operations people tend to warm to it while accounting people frequently turn away. There is a key and crucial role for accountants in moving a company towards World Class Business standards, and activity-based costing is a method to allow the accountancy profession to get heavily involved in de-veloping a company to World Class levels.

Throughout the previous chapters we have looked at wastes — wastes that cost money and destroy profit. Traditionally, the ac-countants in a firm have been the guardians of a company's profit and activity-based costing provides them with an effective tool in a WCM environment.

The current practice in many companies is to attribute mate-rials and labour costs to products to determine their direct costs and then attribute an amount of overhead to the product, usually determined by a cost driver such as volume, man hours etc. This tends to be a simple costing system that is relatively easy to administer. However, very few modern companies operate in such a simple way. They often have quite complicated processes that make it difficult to assign costs effectively to a given product.

Let's take as an example a factory with two main product lines. One product is manufactured using simple, cheap machinery at a medium volume, while the other product is manufactured using a state-of-the-art machine. Both products have similar raw material levels and sales volumes. In a traditional costing system, labour and materials would be costed directly to the products with a proportion of factory overhead based on a chosen cost driver, typically volumes or labour content. The big problem with cost allocation is getting to a suitable level of detail as to where costs should be allocated and in determining what is the most suitable cost driver on which to base the overhead allocation. In the above example the old, cheap machinery could well end up carrying a much more significant amount of the overhead than the new costly machine. The question is complicated by the fact that the new machine may well need significant financing and technical support when compared to the steady, fully depreciated older machinery. Unless the costing system is sufficiently detailed to capture these costs and allocate them correctly, management may well make incorrect business decisions based on less than accurate information.

Activity-based costing stems from a technique known in America as activity-based management (ABM). ABM has a number of similarities to parts of the WCM process. It looks at processes to identify which parts of the process are value-adding and which are non-value-adding. Having identified these separate elements, it then suggests ways to minimise the latter. In this way it complements the aims of World Class Manufacturing.

Activity-based costing differs from traditional costing systems in that it tries to identify and attribute various activity costs — such as maintenance, engineering, support functions etc. — accurately and effectively where they are incurred, rather than bury them in large pools to be distributed across a product range. Reducing the size of the general overhead costs allocation pool gets you closer and closer to directly relating costs to the areas and products that actually incurred them in the first place. Clearly, when management has accurate information it is generally able to make better decisions based on real facts.

To achieve the above, the ABC system looks at costs at different levels:

- Unit Level Costs

- Batch Level Costs

- Product/Process Costs

- Organisational Costs.

Unit Level Costs

Unit level costs are those direct-type of costs such as materials and labour, and specific machine costs such as electricity where this can be measured. The cost per unit is generally significant for each item produced.

Batch Level Costs

Batch level costs relate to cost items that are incurred to produce a given batch, such as the cost of purchasing the components, the time taken to set up the machine or process and get it running properly, the costs associated with inspection and testing, materials handling and the scrap produced during the production process directly. These costs are not usually associated with any given article produced during a batch run, but rather are linked with the total batch.

Product/Process Level

These costs are associated with the next level up from the actual process itself. They relate to such items as engineering changes to the product, the overall costs of maintenance on the machinery, the costs associated with the complete product development process and such items as scraps inherent in the design of the products.

Organisational Level

At the top level of the organisation are the costs of being in business, such as depreciation, administrative costs, support staff and functional departments, marketing etc. They are incurred to allow the business to exist.

CONCLUSION

By focusing on costs in such a detailed way management is able to identify where costs are being incurred, and quite often are able to identify changes in systems and processes that will avoid the necessity of adding to costs or facilitate reductions in them. A useful analogy is shining a torch into dark places. When we can see the costs and where they are being accrued we can do something about them. With a large pool of arbitrarily assigned overheads, the analysis of value for money is often difficult to make. By reducing the complexity of the system, reducing the size of the cost pool and identifying specific costs and cost drivers, management has the opportunity to correctly identify actual costs and take actions to address them. The situation has often arisen in companies where traditional standard accounting practices indicated that a product line was losing money and should be axed. After applying World Class principles to the production process, and using an ABC system to identify potential cost savings, such product lines have often become key earners for the companies.

In this way it is obvious that the costing system chosen can affect behaviour and management decisions in relation to operational systems, marketing and continued product line-up. It is very important to ensure you do not go overboard when putting in a new system. The benefits you can expect to receive need to be quantified, or at least estimated, before embarking on such an exercise. ABC is an accountant's view of the financial side of operational development and thus relates closely to the other tools and techniques of a World Class programme. When the accounting staff of a company become fully integrated into a company's WCM programme, they can make a significant contribution to its success.

Chapter 8

Supplier Development

So far we have looked at the application of WCM principles within the four walls of a company. We have examined what happens on the production floor, in the offices and in the design and marketing departments, to help satisfy if not delight customers. But this only looks at part of the total value chain.

FIGURE 8.1: THE VALUE CHAIN

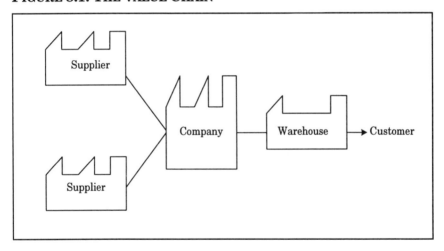

The amount of value added to a product in any factory varies. Studies have shown this value-adding can vary from 20 to 80 per cent, with many of the larger companies adding only 20 per cent if not less to the products they sell. Typically a company will add 30 to 40 per cent to the value of a product passing through its processes. If a company has control over only 30 per cent of the value of its products it becomes very important to try to get some control over the other 70 per cent of its costs.

In the past few years, WCM companies have started to address this issue by developing a special relationship with their suppliers that can help them to develop internally. The large multinational companies have been at the forefront of this activity, demanding quality information such as statistical process control charts and capability study results from their suppliers. Also, the advent of ISO 9000 introduced the idea of vendor audits and continuous vendor rating.

As the WCM concept has evolved and been accepted by the multinational corporations, they have attempted to move more and more of their in-house expertise to their suppliers. In a number of cases this has been done in a rather imperious way, but in the better examples it has proven to be useful for both companies.

PURCHASING — BY PRICE OR COMPETITIVE ADVANTAGE?

The purchasing manager in the West typically holds an adversarial position with the suppliers — suppliers are there to be squeezed, to reduce prices, to supply when demanded. This has led to a well-understood customer–supplier relationship: the supplier keeps the customer at arms' length and does the minimum to keep the customer happy, because anything more costs money. Frequently, suppliers will tender for business at an unsustainable level in the belief that price increases will be negotiated after the first year of doing business, as a level of dependency is built up. This situation puts considerable strain on both the supplier and the purchaser.

Many larger companies have moved away from being primarily manufacturers to being system designers, integrators and assemblers. Ever larger amounts of the value of goods are made up of bought-in components. Henry Ford's idea of an automobile manufacturer taking in coal and iron ore at one end of the factory and converting it to cars at the other end is far removed from today's car manufacturers who are to a large extent assemblers of other people's products and sub-assemblies. Let's take a closer look at a computer manufacturer, as an example.

The major computer companies effectively integrate components and sub-assemblies from a wide range of suppliers to make a working computer. The companies specialise in innovative integration of these disparate elements, defining how components should interact. A typical analysis of their supplier chain shows that the companies themselves only account for 8 per cent to 10 per cent of the cost of the products it produces. Eighty per cent of the delivered cost of a computer is under the control of the suppliers to these companies.

FIGURE 8.2: COMPUTER COMPANY VALUE CHAIN

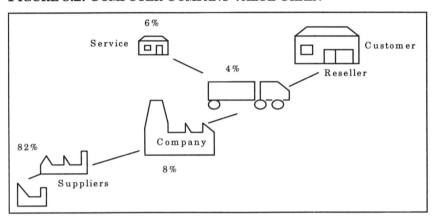

Changes in the marketplace are forcing these companies to examine their complete value chain. Customers are rapidly changing their requirements and the companies are trying hard to interpret these changes and respond effectively to them. In general, the market is demanding:

- Reliable, high quality products
- Exceptional price/performance ratios
- Fast, reliable delivery
- Timely, accurate information
- Value for money.

Customers have shown they would like:

- High-spec, low-cost products
- Software to be pre-installed and delivered with the machine

- Products badged and boxed as a re-sellers own brand

- To have third-party cards installed — enhancing the product configuration

- To be able to specify configuration right down to the basic chip level.

These needs and wants are constantly changing with many of yesterday's wants becoming tomorrow's needs. This puts a great strain on a company's internal systems and suppliers — a strain that has made it critical for Western companies to move away from the traditional adversarial way of doing business with suppliers.

The system in the Far East works differently. The supplier–customer relationship is built up and developed. Initial contract prices are negotiated, often based on target costs. The customer will give feedback and support to the supplier during the course of the contract but will expect the supplier to be making improvements in product processing all the time. The customer will expect to benefit from these savings, will look for price reductions and will often look for additional features when the contract comes up for renewal. The Far Eastern company will tend to rely on its suppliers for new innovations in product processing and in general process development.

The difference between the two systems lies in the area of co-operation. In the West we have been loathe to co-operate, while in the East they have been open to the concept and embraced it. By focusing on the potential competitive advantages available from working with competent, progressive, financially sound and innovative suppliers, a company can optimise its products to take full market advantage of this co-operative effort. The co-operation is not based on any altruistic feelings the companies may have for each other — it is based on the belief that progress and shared profitability is good for both of them. They tend to have a feeling of a shared destiny.

The complexity of our modern world and our modern products makes it very difficult if not impossible for a manufacturer of assembled products to master all the technologies contained in that product. Can an alarm manufacturer be expected to understand

and master the intricacies of computer chip design, or is it better served by focusing its expertise in the area of system integration, chip programming and the innovative provision of features for its product? Can the farm machinery company be expected to master heavy axle manufacture, or is it better served focusing on the machinery to be carried by the axle?

These companies need to have the support of professional, competitive and innovative suppliers to ensure their end products match, if not exceed, the customer requirements for their products.

HOW MANUFACTURING SUPPORTS THIS STRATEGY

The computer industry is relatively new but it has still gone through many of the phases in modern technology. The changing strategic perception of the market and its requirements has led to the identification of a number of areas where manufacturers need to improve to meet market demands. The primary concepts are *speed*, *consistency* and *agility*. To achieve improvements in these areas manufacturers must close the loops between themselves, their suppliers and their customers by communicating effectively. In this way, the disparate processes of a business can be brought together as aligned elements along the value chain, minimising waste and maximising responsiveness to customer needs.

Speed

Manufacturing has focused on cycle time reduction as the driving force for its efforts. By analysing process and physical flows, addressing changeover time reduction and introducing new layouts, cycle time can be reduced from four days to four hours, changeovers can be reduced from 60 minutes to 15 minutes.

Companies have identified wastes in the production area and taken them out using JIT tools and techniques. This allows manufacturing to respond quickly to market demands.

Consistency

But speed without adherence to quality would be useless. There is no point in a supplier getting the parts to your line if they are not of top quality, if they are going to fail in assembly or use. Companies need to work with suppliers to interactively develop the quality levels between them. The use of ISO standards can form the basis for such a development process. Efforts to improve speed of response cannot be allowed to adversely affect the quality of products. The rigours of the ISO 9000 system ensure that companies can be confident that quality targets are being met.

Many companies realise that a passive quality system was not getting close enough to their customers' needs and possibly not getting full feedback from the market. Studies have shown that only 5 per cent of dissatisfied or less than satisfied customers will complain, the other 95 per cent just won't buy your product again! Many companies have introduced a system of proactively working with customers to get quality feedback. Such a system checks that their customers are happy with the product they bought and that the quality of the product was what it should be. This idea of relating with the customer after the sale is not new but it *is* an effective way of caring for the customer and developing customer loyalty. This loyalty results in future sales.

When such quality information received directly from the end user is fed back along the supplier chain there are ever-increasing opportunities to find the root causes of problems. Each supplier along the chain has the opportunity of helping in the process of overall product quality improvement and so securing further work from ever more satisfied customers.

Agility

Making standard products with speed and consistency within the four walls of a factory is not enough for today's customers. Companies need to be able to respond quickly to the varieties of the marketplace. Manufacturing needs to develop its flexibility by allowing for a variety of products to be manufactured with minimal overall disruption. If you need to be agile and you are part of a chain then all the elements of the chain also need to be agile. Unless a supplier takes on board the base concepts of WCM then their agility will come from high stock levels or excessive man-power inputs to meet the changing needs of the customers. This agility can only be achieved at reasonable cost by the mutual sharing of ideas, concepts and information. Agility can be assured by companies working together to synchronise their operations along the value chain.

Closing the Loop

Agility, consistency and speed of response cannot be sustained without close integration across the value chain.

The original JIT model resulted in stable processes in-house and close links with key suppliers. The fall-out from this approach was that products were pushed from stocks into customers. Since market requirements have changed so radically the links in the chain have to be stretched to include the customer.

FIGURE 8.3: ORIGINAL JIT MODEL

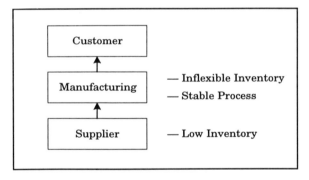

FIGURE 8.4: THE WCM MODEL

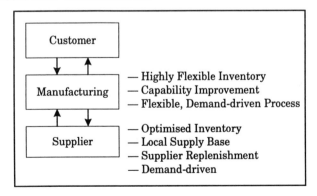

The production cycle is driven totally by the desire to give maximum benefit to the customer. By so doing the customer will be delighted and develop a loyalty to the product. This loyalty ensures the future profitability of the business, while reducing and eliminating wastes from the organisation protects present profits.

Communicate to Survive

Communication is the key to this flexible approach. The manufacturer needs to know quickly what is happening in the marketplace and pass this information along the chain to their suppliers. Figure 8.5 illustrates an example of effective data links.

The lead time in the manufacturing operation has been reduced so significantly that real time information from the marketplace is essential. When a product is sold it can be commissioned for replenishment from the production line rather than from an intermediate store. The link to the supplier becomes critical to ensure the chain is not broken. A local supply base is essential to ensure production can satisfy market requirements. Local does not mean that a supplier plant needs to be located physically close to the plant, just that it can respond quickly to needs. The use of electronic data interchange (EDI) can assist the flow of information both to and from the suppliers and thereby remove further wastes from the supply chain.

FIGURE 8.5: WORLD CLASS DATA LINKS

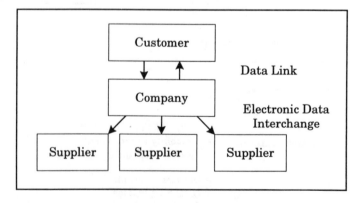

Business in general, and the computer industry in particular, is moving towards the goal of much greater supplier integration. The process will take time but the goal is attainable and worthwhile. The integration will result in a much more streamlined supply chain.

WORLD CLASS CLUSTERS

The idea of supplier development was originated by the major multinational companies. The relationship was often very uneven, with a large company possibly employing thousands telling a typical SME with 60 to 100 employees what it wanted done! The small company would work harder and try to achieve what was required of it — or stop doing the business. The supplier development process took on something of an authoritarian system, with the SME attempting to comply but often not quite understanding why things had to be done that way or fully accepting them. Situations have arisen where a company would supply two other companies with a similar if not identical product. Both customers, being much larger than the SME, would insist on *their* way of performing quality checks on the type of quality system to be used. The bigger companies have dedicated quality departments that rarely leave their own premises, as compared to possibly one or two quality staff in the supplier company. Such demands for conformity to their own systems by the major companies puts additional cost burdens on the SME supplier companies.

The World Class Cluster concept is one where SMEs start to work together to harness the potential for rapid development that was, until recently, the sole preserve of the multinationals. Europe is approaching a truly single market status, where products and services will be able to sell equally well anywhere in the EU. This homogenisation of the market now offers European SMEs a golden opportunity for expansion.

By applying the techniques and concepts of WCM along the supplier chain, these SMEs can use their individual expertise in disparate areas of technology to create and develop innovative products with short lead times, at high quality and feature levels and at relatively cheap costs.

The multinationals and the major Japanese companies have pioneered the WCM concepts. They have determined what works well and what is difficult to implement. This is all information that SMEs can get access to with very little difficulty. The leading SMEs are now in a position to bring the concepts of WCM to a much broader audience. As their sizes are relatively compatible with other SMEs, there is a better opportunity to create a sense of co-operation between companies. This is represented graphically in Figure 8.6. Ideas and trials can be discussed in an effort to determine what techniques and tools are most applicable in any given situation. Cross-fertilisation and brainstorming on an inter-company basis have the potential to greatly reduce wastes between companies, just as we have seen it do in the production and office areas.

FIGURE 8.6: THE SME CHAIN

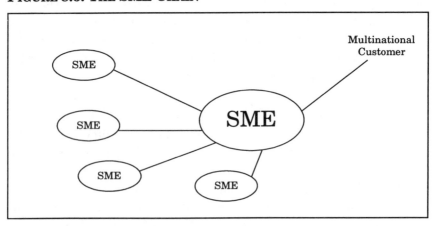

At an earlier stage in the book, the two bin system of parts storage was discussed in relation to cellular manufacture. Figure 8.7 represents the idea of two bin stock control as it relates to a customer and supplier.

The wastes in this arrangement are quite significant. If communication and co-operation between supplier and customer are good and the supplier's performance has been consistent over a period of time, it becomes immediately feasible to remove bin D from the customer's stores. As forecasting improves and as production systems in the supplier's facility also improve, it should become possible to remove bin C from the customer stores. As the overall level of trust and system capability develop the supplier will be able to rely on its production process to satisfy the customer's call-off within the required lead time, allowing it to re-examine the stocking quantities in bin E.

FIGURE 8.7: TWO BIN SYSTEM — SUPPLIER TO CUSTOMER

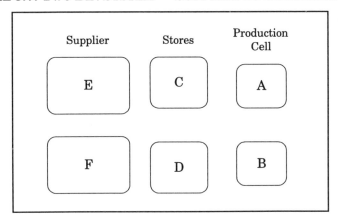

This system is different from the original interpretation of JIT from the early 1980s. Then the system was understood — or at least it was the practice — that the supplier would take bin D and most of bin C back into its stock, allowing the customer to run their stock holdings towards zero. The cost burden of holding stock was pushed back onto the supplier. The WCM approach focuses on reducing the necessity for major stock holdings right along the value chain by addressing the reasons why such high stocks were necessary in the first place, by:

- Reducing processing lead times

- Improving communications

- Identifying and optimising key processes.

As SMEs take on the challenge of WCM they become better able to compete with the best. They are positioned to attack market niches with the high level of flexibility they posses and the speed with which they can make changes.

In the example described above, quality and engineering staff, designers and purchasing people would discuss the areas of mutual benefit between the two teams of people, and ask such questions as how to:

- Get the best quality at source

- Use the best processes and systems

- Minimise the cost of acquisition

- Optimise the flow of materials

- Minimise paperwork between companies.

The basic ideas of WCM — *Quicker, Better and Cheaper . . . Together* — have a valid application in the supplier–customer relationship. As the benefits of WCM come through they can usually be shared. If a company is better able to compete in a market it is likely to sell more product, which in turn requires more parts which can be produced more profitably by the cluster companies than by competitors who stick to the traditional arms' length-type of customer–supplier arrangements.

Chapter 9

World Class Manufacturing and Company Strategy

Throughout this book we have dealt with the practicalities of World Class Manufacturing. We have looked at specific tools and techniques that can be implemented in a systematic way in companies of all sizes. It is possible to make major improvements by addressing these operational issues, but it is also important to be aware of a number of strategic issues before starting the process. After all, meeting the strategic objectives of a company is a major reason for the enthusiastic interest in World Class Manufacturing in the first place!

This chapter is not intended to be a guide to strategy and its formulation; there are numerous books available totally dedicated to the subject. Instead, this chapter will primarily focus on how the top levels of strategic planning are effectively supported at the operational level by World Class Manufacturing.

STRATEGIC PLANNING

Every company has to define its relationship with its markets and its competitors — it needs to know what it does and where it is going. A strategy defines this relationship, and the message is then transferred to everybody in the company in the form of a Mission Statement. The goals of the company are reflected in the Mission Statement through a number of measurable objectives to be achieved.

In a small business, this strategy formulation process often takes place within the head of the owner-manager, it is their mission, their plan. Quite often, the plan is not written down, and is only obvious by the activities the company undertakes. In larger companies, more formal strategic plans are necessary in order to focus all the staff on the same objectives and to ensure that everyone knows where the company is going.

The strategic planning process should be a continuous one. Senior management need to be aware of the changing environment the company operates in. This is especially true in a World Class company. For example, what would have happened to a World Class music record manufacturing company if they had not been aware of the advent of the new technology of the cassette or compact disc? The strategists within a company need to keep in touch with such technological advances or major changes in the external business environment. The strategists also need to keep in touch with the changing needs of their customers and of their competition.

When a World Class company starts to formulate its strategy it should have the philosophy of World Class Manufacturing as its guiding principles. It needs to be fully aware of the necessity to be:

- Flexible

- Responsive

- Competitive

- Innovative

- Quality conscious

- Customer aware.

This book has shown that companies need to be "quicker, better and cheaper" in their response to customer needs than their competitors. Another key point of the World Class philosophy is the "together" part of the equation. In an operational context, "together" relates to all members of a company working effectively as a team. At a strategic level, it relates to the integration of the disparate elements of the company strategy to best satisfy customers' needs. Applying World Class philosophies to a company brings the different departments together so that the strategies of individual departments are formulated with a view to the total targets for the company. Integration and a single focus for all is the goal.

Successful World Class companies have achieved a high degree of success in integrating the business, marketing, manufacturing and financial strategies and aligning these with the human resources needs of their companies. At the same time, they are setting the pace and standards for all other companies worldwide. A World Class company may be located around the corner or on the far side of the world. But unless you take them seriously, you are likely to lose business to these new competitors.

LINKING BUSINESS FUNCTIONS AND STRATEGY

Manufacturing

Linking business and operational strategy is a central building block of a World Class company. Decisions made when designing and running the production system need to be consistent with the overall strategy of the company. If the decisions are unrelated or inconsistent they will affect productivity negatively. If the company wants to become World Class but does not take on board the principles of Just-in-Time Manufacturing, or is unable to implement an effective quality programme, then there is little hope of success in an increasingly competitive marketplace. In this way, decisions as to stock holdings, automation, schedules, staff involvement, and so on need to be made with regard to overall World Class principles.

Marketing

Consideration must also be given to how the company's marketing strategy fits in with the overall business plan. The focus of marketing is on how best to serve the customer, while at the same time making a profit. As a result, the marketing department must be close to the customer while continually monitoring its competition. It is the area best positioned to understand the firm's competitive situation and to relay this to the rest of the organisation.

In the WCM company, the full team must realise that the marketing department cannot formulate its strategy in isolation, that marketing plans need to be interwoven to the overall company strategy. In a more traditional company it is assumed that all a company needs is a marketing strategy that gives it market orientation and so provide the market with what it wants. This is no longer enough. The marketing department needs to identify where the company will find its market niche, what its competitive position will be, how products will be planned, the required pricing and the methods of distribution. Manufacturing strategy must support the particular thrusts of the marketing strategy. The World Class concept works to bring these diverse elements together to best achieve the required market demands of improved quality, cost and lead time. The World Class company is able to respond quickly, flexibly and effectively to changing market demands, not by working harder but by working smarter and designing its systems to allow and facilitate this level of effective responsiveness.

Human Resources

A key element of any company's strategy is its human resource strategy. It is also a key element of the World Class philosophy. The ability of a company to address the changing demands of a market is completely dependent on the ability of its staff to cope with these changes. Without this ability to deal with change it is practically impossible for the manufacturing function to support any strategic initiative. A company's people make the strategies work, they carry out the plans and achieve the results. It is by harnessing the abilities of staff that competitiveness can be achieved and improved upon.

Companies implementing World Class systems give a great deal of attention to people-related aspects of the workplace. The original idea of the personnel function being concerned solely with administration of employees is gradually being replaced by a much broader concept involving the staff in the overall development of the business. This viewpoint takes in a much wider range of human behaviour in organisations. The company needs to address ideas concerning changes in culture and structure, the distribution of power and the handling of conflicts inside the organisation.

The World Class philosophy assumes everybody is part of the solution, that their contributions are valued and are more likely to be constructive than would be the case in a traditional company. The philosophy also accepts that human behaviour is extremely complex and difficult to understand in a systematic way. Therefore, it does not try to systemise the personnel function, rather it endeavours to take the adversarial part out of the interaction by incorporating people into the problem-solving process. If the people in a company are not aligned to its manufacturing and overall business strategy then even the best plans of management are likely to fail.

Finance

The last element of strategy that forms an integral part of the World Class philosophy is in the area of finance. Decisions taken in the other functional areas clearly have a financial dimension. Manufacturing, for example, usually ties up the vast bulk of a company's assets, so before investments or organisational decisions are taken the financial implications need to be understood. The financial function is primarily focused on the acquisition and deployment of funds, financing the business and measuring and managing the flow of money into, around and out of the company in the course of business. The World Class philosophy requires the financial function to contribute significantly to the identification of wastes throughout the operation, not only in the production area but also on the administrative side of the business.

MANUFACTURING AS A COMPETITIVE WEAPON

The use of manufacturing as a competitive weapon is now an accepted norm by many of the world's top companies. It should be recognised that a production system can be either a support to the strategic goals of a company or a dead weight holding it back. In the traditional company, top management has often abdicated responsibility for large portions of business strategy to operations management. As a result, many companies are often encumbered with an inappropriate resource mix to achieve the targets required of them. Once acquired, these resources can be very difficult to replace with the right ones, and assets can quickly become liabilities.

In the World Class company manufacturing takes on a strategic importance. The new approaches of Japanese companies, with their long-term investment outlook, market growth emphasis, skilful financing and focus on being a low-cost, high-quality producer, are seen to be those of the twenty-first century. It is obvious that manufacturing management is faced with a new challenge — to be able to produce products Quicker, Better and Cheaper. To do this companies have to move away from the mass production concept to one of mass customisation or craftsmanship, to embrace the concepts of World Class manufacturing to achieve these goals.

The success of the World Class Manufacturing strategy pursued by Japanese and leading Western companies provided the basis for the examination and interpretation of this new approach to manufacturing presented in this book. The approach is not restricted to multinationals with large resources, however. The simplified tools and techniques described in this book will allow small to medium-sized companies to deliver constant improvements in productivity and quality on their way to becoming World Class companies.

Appendix 1

What About the Japanese?

The leading Japanese companies have moved through the stages of Quality Systems, Total Quality Control, Employee Involvement, Waste Elimination and Early Supplier Development to the start of the World Class Business Development process.

The Japanese have moved from a position in the 1960s where their products were perceived as poor quality and cheap, to the present day where they dominate many sectors of the world market, holding over 50 per cent of world export trade in many of today's most important products.

FIGURE A1.1: PRODUCTS WHERE JAPAN HOLDS OVER 50 PER CENT OF WORLD EXPORT TRADE

• NC machining centres	• Cathode ray tubes
• Hi-Fi systems	• Tape recorders
• Video cassette recorders	• Pianos
• Copiers	• Magnetic tape
• Cameras	• Electronic calculators
• Microwave ovens	• Electronic typewriters
• Shipbuilding	• Ground satellite stations
• Liquid crystal displays	• CD players, etc.

Source: Hines (1994), *Creating World Class Suppliers*, p.31

One of the consequences of this rapid rise in fortunes for Japan has been the equally rapid rise in the value of its currency. This rise has put severe pressure on Japanese manufacturers to sustain competitiveness. Wage rates in Japan are high so local manufacturers have had to strive to achieve savings in the order of 30 per cent on manufacturing costs, this from a base already considered low by Western standards. They have achieved these savings through the use of a number of devices, in particular *Kyoroyku-kai* or supplier associations, *Kaizen* or continuous improvement and increasingly through the use of Total Productive Maintenance or TPM, which is a Japanese equivalent of WCM, particularly suited to large companies.

The major companies and corporations are also sourcing more components and sub-assemblies overseas, particularly in South-East Asia and China, where they are also locating overseas operations of their own. These often take the form of joint ventures where they can benefit from the low labour cost structures of their neighbours. Possibly the most interesting developments are being achieved in companies who decide to sustain their operations in Japan itself. These companies need to achieve the same levels of savings, circa 30 per cent, and are having to fundamentally address their internal processes, re-examining their product designs, manufacturing operations and sales/distribution systems. The pressure brought to bear to achieve savings of this level is quite severe and demands the extensive utilisation of the tools of WCM to achieve them. These pressures are causing major changes within the overall Japanese business system, with companies such as Toyota starting to source components outside their traditional *keretsu* or company group. One of the "new" tools developed to help achieve these cost reductions is that of Target Cost Management. This is simply explained by the following two equations:

Japan: Sales − Profit Required = Target Manufacturing Cost

Western: Manufacturing Cost + Profit Required = Sales Price

By putting the pressure on the design and manufacturing divisions, the most effective Japanese corporations are remaining competitive in the face of severe currency appreciation. But this is not being achieved without pain, and often severe pain. Nissan have only recently returned to profitability after three years of staggering losses. They had to take severe action and closed a major manufacturing plant in Japan as part of their restructuring, an action that has been largely unheard of since the end of the 1950s, and a major shock to the Japanese system. Japanese managers are now having to face some of the personnel pressures that their Western counterparts have come to terms with over the past twenty years. It is significant however that this closing of the plant was seen as a final measure and not simply as a way of increasing overall corporate profitability.

Japanese companies have effectively honed and refined the basic concepts of WCM and are applying them to aspects of their businesses where they have not been traditionally acceptable. Many of these concepts can be grouped under the continuous improvement phase of a World Class company.

Major advantages to be obtained in this phase of development appear to reside in supplier development or, more accurately, strategic supplier alliances. As companies come to rely more and more on their supplier's experience and expertise, the question of supplier capability and performance become ever more important.

KYOROYKU-KAI

When Toyota, Nissan and Hitachi first developed their ideas in relation to TQM and JIT they were faced with a problem we never had to face in the West — nobody knew what they were talking about! They had large numbers of suppliers who did not understand what they meant when they talked about quality tools or JIT techniques. After a group of small to medium suppliers to Toyota came together to share resources and information, it took Toyota four years before they formalised the system and developed it to incorporate more and more of its key suppliers into Supplier Associations, or *Kyoroyku-kai*.

The *Kyoroyku-kai* arrangement can be illustrated as shown in Figure A1.1.

Companies such as Toyota found that this system of supports was an extremely effective way of getting ideas and concepts passed along the value chain. If they took the time to train and develop their first tier suppliers they would in turn develop their suppliers and new concepts, ideas and techniques would spread quickly. The idea has found its most widespread acceptance in the industries where the bought-in content is high and the company relies greatly on its suppliers for competitive advantage.

FIGURE A1.2: *KYOROYKU-KAI* **SUPPLIER TIERS**

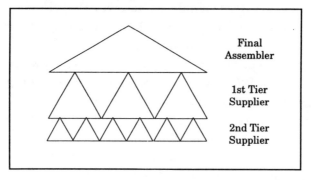

Final Assembler

1st Tier Supplier

2nd Tier Supplier

BEST PRACTICE — *KYOROYKU-KAI* IN JAPAN

The top level associations meet, with the managing directors and possibly senior managers discussing overall market trends, new technologies and the strategic plan of the lead company. In this way companies can be sure that their independent strategies are complementary. These meetings would take place once or twice a year, with the senior managers setting policy on the agreed strategies for their operational staff.

Workshops are run for the next level of managers where the problems associated with implementing the policies decided upon are discussed and resolved. These workshop sessions are frequently used for the transfer of tools and techniques and form a core part of the workings of the system. As the associations consist of suppliers to the major company, there is a feeling of shared destiny among the association members. If the major company prospers then they in turn will prosper. Developments and enhancements of tools and techniques are passed freely among association members, for the betterment of all.

Factory visits are organised to allow staff to see environments other than their own. The Japanese are very partial to the "social event" and these form an integral part of the associations team building and cohesion creating activities.

In Japan, 50 per cent of the GNP is produced by companies employing less than 100 people. The importance of distributing information and knowledge to sub-suppliers of the major corporations is critical to the continued success of these companies. The main objectives of the supplier associations is to foster an in-depth understanding of JIT, TQM, EI, SPC, CAD/CAM, Management Flexibility and Cost Reduction concepts right along the supplier chain. As suppliers develop and evolve, the flow of information and ideas develops into a bi-directional flow also helping to strengthen bonds and communication levels between the supplier and the customer with a view to forming true strategic alliances. When both supplier and customer are using the same tools and speaking the same language, inter-company wastes can be identified easily and eradicated.

The two-way communication allows for both companies to keep in touch with market trends and shifts, to keep close to the customer. The supplier association arrangement can be compared to the Roman Legion system, where the optimum number of men reported to an officer and so forth up the line of command. In the industrial arena, it is very difficult for a purchasing department and an engineering department of a company to develop and foster relationships with a large number of companies. The traditional approach is presented in Figure A1.3.

The Supplier Association approach looks more like the arrangement presented in Figure A1.4.

FIGURE A1.3: TRADITIONAL SUPPLIER ARRANGEMENT

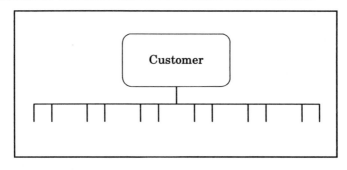

FIGURE A1.4: *KYOROYKU-KAI* **CUSTOMER–SUPPLIER ARRANGEMENT**

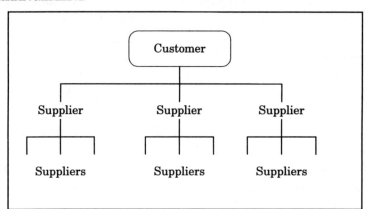

KAIZEN

Kaizen is another Japanese word with a simple meaning — Continuous Improvement. The Japanese believe that it is possible to do things better and that this can be achieved a little bit at a time. Many companies in the West don't focus on making little improvements, in the belief that these little improvements are too small to worry about. The Japanese concentrate on these small issues and by amassing a large number of them achieve significant improvements in overall performance. This different approach to improvements and developments can have a significant impact on levels of competitiveness. Let's look at two companies:

Company A is a traditional company. Its technical staff buy good quality, up-to-the minute machinery, which they maintain and operate correctly. They are up-to-date! Well, at least as up-to-date as the manufacturers. The only problem is, how often have machine manufacturers actually run machinery in a production situation? (see Figure A1.5)

Company B is a World Class company. Its technical staff keep a constant watch on what is happening in the new machinery markets. They analyse what new additions and features have been added by the manufacturers. They ask themselves and their operations staff if these features could help them. They integrate closely with production staff and know how to run production machines. They help to actively solve problems on machines and improve their operation. The company's machines are often fitted with non-standard parts and features designed to improve productivity by making jobs easier, and thus it is more difficult to make mistakes. When they buy new machinery they ask for their own modifications and improvements to be incorporated. (Note: This can be a problem for Western companies buying machinery from Japan. The Japanese suppliers expect the purchaser to know what they want the machine to do!) These additional features allow the World Class company to differentiate itself and its processes from the competition (see Figure A1.6).

FIGURE A1.5: INNOVATION WITHOUT *KAIZEN* (COMPANY A)

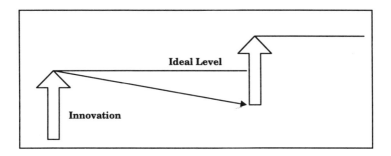

FIGURE A1.6: INNOVATION WITH *KAIZEN* (COMPANY B)

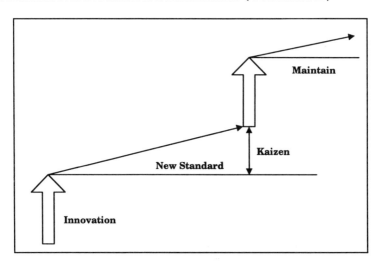

So what are the competitive differences between companies A and B? Company A buys the most up-to-date machinery and effectively competes on the basis of good management and effective cost control. Company B buys machinery with its own modifications and competes on the basis of good management, effective cost control and superior production processes, achieved through its internal developments of machinery and systems of production. It also has less capital invested in expensive machinery and has therefore a better working capital position. These improvements are typically based on years of in-house understanding and expertise.

If another company wants to enter the market place it has to compete with companies A and B. To compete against company A it needs to buy the same machinery and run it effectively. To compete against company B it needs to identify the most appropriate machinery and develop it and its system of supports to a high degree. The problem of competing with company B is clearly much more difficult. The process of Continuous Improvement of machinery and processes allows a company an opportunity to take a strategically advantageous position within its markets, a position it is difficult to overcome. The use of Continuous Improvement/ *Kaizen* process developments allows a company the opportunity to remain competitive in the face of increasing competition from cheap labour cost countries.

PRODUCTIVE MAINTENANCE ORGANISATION (JIPM)

The major Japanese companies have been the world leaders in the areas of JIT, TQM and Employee Involvement. Companies such as Toyota have led the way and spread an understanding of the concepts through their supplier associations or *Kyoroyku-kai*. This system worked for the major companies and their associated suppliers, but not for the many more companies that were not actually associated with the leader companies. A state-sponsored organisation, JIPM, was created as a means of spreading the messages of Total Productive Maintenance incorporating Just-in-Time, Total Quality Management and Employee Involvement throughout these companies.

The organisation provides a mechanism for training and accrediting companies in the concepts of TPM. There are two levels of award that effectively grade a company. An integral part of a company's evaluation process examines how well it helps other companies to develop their understanding and use of the TPM (WCM) tools. In this way the message of TPM is spread throughout a wider body of companies. The emphasis is strongly put on training and dissemination of the underlying concepts. They believe that a high level of training and a development of core understanding of TPM concepts provides a high entry level barrier to competitors which thereby strategically supports companies and the economy in general.

If a company has to train and help others to understand the principles of JIT, TQM and EI, it has to know them well. This method ensures that ideas and concepts are well understood along the supply chain. As a company and its suppliers develop their mutual understanding of the TPM concepts within their individual environments they also begin to understand that they have applicability between their organisations. They begin to focus on the interactions between their respective companies and effectively address issues between them. In this way, wastes along the supply chain are identified and eradicated.

OVERVIEW

Close analysis of the Supplier Associations of major Japanese companies indicate that each of the major firms has its own approach to the idea. They have interpreted the concepts and applied them in a way suited to their own company visions and cultures. Some associations are run on very open, democratic lines, while others are run in a paternalistic, authoritarian way. The level and type of involvement of the major companies in their supplier companies can vary from being quite significant to the point where the supplier companies are driving the big company to act.

Companies are asked to participate in the Supplier Associations based on their ability to contribute, their importance to the big company and also their willingness to be part of a strategic alliance with the big company and the other members of the association. Not all these factors have equal relevance in the West — it is up to us to interpret for ourselves a suitable understanding of the concept that will work for us.

Appendix 2

Case Studies

Not all WCM projects take place in "new industries", in electronics or plastics or the automotive industries. Not all projects follow the same route, or are carried out in profitable, stable, forward-thinking companies. These two case studies are taken from companies operating at different extremes of the business spectrum in terms of management style, products, processes and profitability.

TANCO ENGINEERING COMPANY LTD

Tanco Engineering is located in Bagenalstown, County Carlow, in the Republic of Ireland. The company has been trading for over 30 years and was, at the start of the WCM initiative, still led by its founding directors. The company designs and manufactures a range of agricultural machinery from sophisticated silage bale wrappers to hydraulic front-end loaders. Most of the company management had no work experience other than within the company.

The company sells its products in over 20 countries worldwide and has a reputation for making strong, good quality machines. The silage bale wrappers were regarded as being at the forefront of the technology, in their day. The company invests heavily in research and product development and has an on-going commitment to this aspect of the business.

The company had experienced severe financial and trading difficulties over the three years before starting the WCM project and this was one of the primary reasons why the management team approached the concepts of World Class Manufacturing. The company employs over 70 people and is unionised. Management style at the start of the WCM project could best be described as being very traditional.

Key Problem Identification

The company first became aware of WCM in May 1994 when the financial controller sat through a basic presentation on one of my roadshows. The difficulties associated with implementing a WCM programme in a traditional company were outlined very clearly. Management were advised to learn about WCM before the official start of the programme, which they did by having a number of presentations made to the management team and by reading books on WCM in an effort to develop an understanding of its concepts and tools. The majority of this study was carried out by staff in the middle and lower middle management areas rather than at the very top level.

The diagnostic phase started with a factory tour where problems of high WIP levels, difficult to monitor and control production, mixed production and cramped conditions were evident. Production management was initially unwilling to see the difficulties in the plant, but at least they were prepared to give the concepts a try. The company employed two staff full-time to progress chase batches through the production process.

A World Class Manufacturing team was formed consisting of the senior and middle management of the company. The team discussed in a very open manner and identified the main problems facing the company. The scale of the operation, the range of products and the seasonality of particular products led to the selection of the Autowrap Silage bale wrapper as a pilot project for the development of an understanding of WCM as it applied to Tanco. This diagnostic phase was started in August 1994, with the Autowrap season starting in January 1995, allowing a three-to-four month period to develop this understanding of WCM and to prepare for the manufacturing season. The company felt that it had missed a significant number of sales at the end of the 1994 season. The key findings of the analysis were:

- Product range needed to be rationalised — too many models

- Optional equipment needed to be rationalised — too many options

- Sales forecasts to be prepared, by model, by market, by delivery period

- Sales/Marketing plan to be agreed as the basis for other departments to plan on

- Production and purchasing savings were required

- Manufacturing savings were required versus standard costings.

Parts were seen to pass from one side of the production floor to the other as they moved through the production process. The WCM team used the Process/Physical Flow tools to identify areas for potential improvement.

The company decided to address these issues as part of the WCM programme.

WCM Implementation

WCM Team — The WCM team was formed from representatives from all the major departments within the plant and addressed itself to the question of product range rationalisation and forecasting of sales. A Pareto analysis of the sales of products by market for the previous season was carried out. This led to the reduction of the round bale wrappers models on offer from 28 to 12 models and of the square bale wrapper from 12 models to 5. A similar analysis and rationalisation of the *options* available on these rationalised models resulted in a significant reduction in the number of options available. The sales team then focused on this reduced model line up and proceeded to sell these models, in an effort to get a firm understanding of and basis for the forecasts for the products. The fact that the sales team had to forecast, and that this forecast would form the basis for the other department's plans forward, helped to focus their minds on the task. This focus resulted in the first two months of the season being pre-sold by the end of November, with old stock from the previous season also being cleared out. The creation of the agreed sales plan allowed the other team members to prepare professionally for the 1996 season.

Purchasing — The availability of a sales forecast enabled the purchasing department to develop supplier relationships which in turn resulted in improved terms of trading, better deliveries and enhanced specifications without price increases. The sales forecast allowed the purchasing department to schedule deliveries to suit both internal production and suppliers' manufacturing schedules. Deliveries were actively moved towards Just-in-Time, with KANBAN based supplier relationships developed with local suppliers; parts arrived on time to meet scheduled production plans. The number of suppliers was reduced with long-time problem suppliers identified and dropped. Local intradepartmental teams were created to identify problems and provide solutions to them, within and across departmental boundaries. As problems were identified between departments there now existed a means and a willingness to address them.

Stores — The WCM programme allowed management to focus on the materials as a support to production. Systems were developed internally, allowing obsolescent parts to be identified. Cycle counting was initiated along with an ABC analysis of parts, ensuring a close monitoring of class A items, thereby contributing to working capital savings.

Production — A presentation was made to the full production staff on World Class Manufacturing. Teams were created within the main operational areas and the layouts and flows of material were discussed openly between operators, supervisors and management. A number of small teams of about 5 to 7 members were formed in the key areas of machining, welding and final assembly. The movement of parts through the plant was examined by these teams with a view to identifying if these wasteful movements could be removed. The plant had been laid out in a traditional departmental way, with all drilling machines located in one area, all bending in another etc. The teams started their work in a meeting room, away from the shop floor. This posed a number of problems, as it was felt that the interaction between all concerned was not very good. When the meetings were moved to the shop floor the participation of the shop floor staff improved immediately, probably because they were now working in an environment that was real and were actually looking at the problems face-to-face.

A consensus was quickly reached on both a Stage 1 and Stage 2 layout developments. The Stage 1 layout was achieved over the Christmas holidays by re-arranging machines into a more product-oriented form. Machines were arranged to facilitate the production process rather than a functional one. This meant that a number of machines were turned to make materials flow easier, while others were physically moved to a position where workpieces for plate drilling were located beside the profile machine rather than in the drilling area. Stage 2 involved a more serious movement of plant. One machine weighed nearly 30 tons and had been rebated into the floor. This machine was also one of the most difficult machines to change over from one product to another and frequently required such changes. A replacement was found for this press and the Stage 2 layout was carried out during the following Summer holidays.

The early designs for Stage 1 were displayed in the general factory area and staff were asked for any suggestions. Management was very pleasantly surprised when the suggestions that came from the floor very closely resembled their early thoughts as to how Stage 2 would look. The team concept was and is being further developed to incorporate specific areas such as assembly, machining and welding. A series of small improvement teams were formed throughout the plant with representatives from the relevant sections. These teams focused on local problems such as quick changeover of brake presses, welding jig development, assembly operations and machine operational efficiency. These projects introduced a basic level of TPM to the shop floor with operators being actively involved in improving their machines and/or returning them to their top specification. In the assembly area the total layout and arrangement of the area was re-engineered with consequent significant improvements in performance. The company was using a bonus scheme at the time, and the specific improvements from their WCM activities meant that a number of staff were finally able to achieve a bonus. Video recording equipment was used on a number of occasions to help in the process of identifying possible savings in the operations. The workers themselves were given the copies of the videos and were then integrated in the process of improving their work. Care was taken when using this equipment to ensure that the recordings would not be perceived as being part of a traditional time-and-motion work study.

Present Situation and Plans Forward

The WCM process is on going with inter-departmental communication and co-operation developing. Teams have proved very effective, particularly across departmental boundaries. The company has applied competitive analysis techniques to both processes and products of its competitors and other benchmark companies. They visit other sites to see best practice. Design and production now co-operate in the design for manufacture of new products which is proving extremely effective. The company is working hard to further improve its internal operations by reducing wastes. A WCM Stage 2 programme has been developed and is under way. The topics of this programme are listed in Figure A2.1.

Problems Encountered

The company was in a delicate financial position for the major part of the WCM project to date. Severe financial pressures were applied to the management and this in turn ensured that the WCM initiative was seen as a means of potentially securing the business. The move by senior management to introduce WCM was treated with a high degree of scepticism by the workforce at both management and worker levels. A number of consultants had passed through the plant over the years without seemingly affecting the management thought structures in a significant way. This belief that there was not going to be any change in the traditional approach of the top management team was a major difficulty at the start of the project. The key point followed during the process was that the WCM initiative could possibly hold out a chance for the survival of the company. The middle management and the workers decided to give it a chance to see if the process could work.

FIGURE A2.1: STAGE 2: WCM PLAN

Production

- Development of production cells, including
  ◊ cell team development
  ◊ KANBAN system in production area
  ◊ Introduction of TQM tools
  ◊ Quick changeover introduction
  ◊ Process developments
  ◊ Tooling developments
- ISO system required

Materials Acquisition

- Supplier development programme
- Parts acquisition system
- Parts management and classification system
- KANBAN and two bin systems to be developed

Research and Development

- Product development cycle
- Design for manufacture
- Value engineering
- Teamwork development with production staff

Sales and Marketing

- Integration into WCM teams
- Competitive analysis
- Forecasting and planning with production and materials acquisition staff
- Rationalisation of other products

Administration

- Integration into WCM teams
- Key process development

Over the first 18 months of the process a number of minor confrontations occurred between the senior management, the middle management and the consultant. These typically revolved around the empowerment of people to get jobs done without recourse to senior management for all decisions. The senior management were in effect coming to terms with a serious level of cultural change under very difficult circumstances. The confrontations obviously acted as a deterrent to further efforts but all concerned eventually came to terms with the process.

The financial situation of the company was very severe during this time. Great delicacy was evidenced in the company's efforts to develop suppliers and their interactions with the company during this time.

The workforce were quite well disposed to the WCM effort from the outset. A number of efforts were made to integrate the workers into project teams. These met with a limited response at the earlier stages of the project, when meetings took place inside the office environment. The level of interaction and involvement of the workforce increased dramatically when the meetings were transferred to the manufacturing area. This improvement may well have been attributable to the fact that the workers were now operating in an environment they felt comfortable in as well as the fact that the problems were now tangible and visible rather than conceptual. The difficulty of operating the meetings within the noise and action of the manufacturing plant were more than compensated for by the effectiveness of the workers' participation.

Results

The following key results are specifically attributable to the WCM initiative:

General Comments

- Management team being formed into an effective unit

- Reductions in WIP levels of £30,000, business grown by 20 per cent in the same period, equivalent to an £80,000 reduction in WIP

- Reductions in WIP levels have allowed production cells to be located in areas previously occupied by WIP

- 50 Autowrap units produced and shipped in January 1995. Normally shipments do not start until the last week of January when 7–10 units would be shipped (Company year ends at the end of January, which is extremely significant in terms of end-of-year results)

- Company now making a profit

- The management team are actively developing plans for future seasons of Autowrappers

- The WCM concepts are being applied, independently of the consultant, by the management team across other parts of the company

- Teamwork is developing within the company

Marketing and Sales

- Markets up significantly on previous season to end of July (end of main season)

- Some difficult markets with strong competition from very high spec./capability/price machines, e.g. Germany, Switzerland

- WCM led to rationalisation of models from 28 to 12, with no negative impact on sales

Purchasing

- Good forecasts led to supplier development, good schedules and prices

- 180 suppliers reduced to 100

- Overseas acquisition transport costs reduced by 14 per cent

- Zero stock outs as seen from the production floor

Stores

- Reduction in kitting time from 6 days to 3 days
- KANBAN system set up, saving 2–3 hours per batch
- Forecast allowed potential shortages to be identified early enough to allow remedial action to be taken
- Batches NOT being stopped due to shortages
- Domestic transport down by 60 per cent
- Picking down by 50 per cent; Overtime down by 20 per cent
- Productivity up 60 per cent on previous year
- Faulty warranty parts being cycled quickly
- Cross-functional team set up between Finance, Production, Purchasing and Stores

Production

- Production layout streamlined and focused, parts move through the system; a production progress chaser has been removed from the indirect payroll
- Production efficiency improved. Planned-for production based on standard times exceeded by 25 per cent, on flat time.
- Overtime down by 42 per cent
- Delivery schedule adherence greatly improved, no longer a problem
- 17 per cent reduction in Work in Progress
- Lead times down by 50 per cent
- Cells have allowed a significant reduction in sub-contract costs
- Teams in place with high degree of shop floor participation

Costings

- Accurate costings available for sales team now, 3 months earlier than before

- Labour costs down

- Some materials costs up very sharply but these cost increases can be held due to internally achieved improvements

R&D

- WCM design reduced complexity, part count and assembly time

- Vastly reduced parts count led to easier assembly of machines

- Added features for improved sales appeal

- Closer liaison with sales and production staff

- Integrating field operation results into design

The company is actively proceeding with its WCM initiative. It is further developing the use of teams to improve products and processes on the production floor, in the design area, in the change-over of tooling and the creation of bespoke automated equipment to address specific production problems and to create internal savings. The company was recently purchased by an external investor. The original senior management have now retired from the company.

TRULIFE LIMITED

Company Profile

Trulife Limited designs, manufactures and markets a range of medical prostheses from its Dublin base. The main product is a range of external breast prostheses. In the late 1980s the company developed and launched a range of silicone filled products, leading the field with this innovation. The company suffered financially at the time and was bought by its present owner. They survived this period by focusing on selling existing products and moving away temporarily from advanced research.

The company holds the third position in the world market where there are six main players. Ninety-eight per cent of their production is exported. In 1991 the premises were renovated, and in 1992 they achieved registration to ISO 9002. The quality standard was achieved without recourse to external assistance, with the management feeling it benefited greatly from the experience. Large amounts of production equipment used in the factory have been designed and built internally, and the basic process of making the products was created by staff themselves.

Currently employment levels are approaching 70 staff with a dedicated research and development function. The business is unionised and is run by a young, energetic and enthusiastic management team.

Key Problem Identification

The diagnostic phase quickly focused on the production area and its supports as being the most critical for the company. The production process was analysed to identify problem areas. Key issues identified focused on the level of process control and development within the manufacturing operation. The process consists of making bags, filling them, moulding and curing them and packing the finished product. Key problems identified in each area were:

Bag Sealing

Significant problems were identified with poor sealing of the bags. The level of understanding of the bag sealing process was found to be quite low. Particular areas identified for further research were:

- Application of catalyst

- Sealing conditions — time, temperature, pressure

- Quick tool changeover programme, with up to 30 changes per day at 8 minutes each led to possibly 4 hours production being lost per day.

Filling Area

The bags were filled from a two pot mixer system. The arrangement necessitated long delays when batch ratios needed to be changed as well as elaborate cleaning procedures at the end of the day's production taking up to a man-hour to complete. The filling system also allowed the possibility for air to be entrained into the products, necessitating the placing of the products into a vacuum chamber for 45 minutes to remove entrained air.

Moulding and Curing

Poor quality was recorded from the moulding area, with a significant number of "no nipple" problems being encountered. No rigorous efforts had been expended to determine causes for poor quality with the process not quite understood. The moulding system itself required very intensive manual work with a large number of parts to be assembled onto a complicated mould.

Packaging

The packaging for the product was quite elaborate with the individual packaging for each product size and with a significant amount of labels to be attached to each product. Packaging for about 15 per cent of production was sourced in the USA.

Production Support

The MIS system was felt to be operating under severe pressure, with a number of products being planned manually outside the system. A high degree of double entry of data was evident with consequent delays in getting information to production allied with the possibility for errors in data entry. Internal systems to cover purchase orders (POs) etc. were felt to be less than satisfactory. The production management were under-resourced in the area of process control and development. The main bag material was due for de-listing by the supplier with the consequent necessity to quickly identify and trial a suitable alternative.

WCM Implemention

A core WCM team was formed at middle management level with the General Manager. The decision was taken to address the issues identified during the diagnostic phase across the company. Particular sub-teams were created to address inter-departmental difficulties and these teams reported back to the core WCM team.

The intervention took the format of a series of in-depth discussions with the core team, identifying specific areas of concern and reaching a consensus on the possible solutions to these concerns. Through these discussion sessions the following major issues were addressed:

- Process development engineer appointed — focused on key areas such as bag sealing, filling and moulding

- Quick changeover project on bag-making machine

- In-line filling/mixing machine was identified, trialled and sourced

- Moulding process developments

- MIS — simplified system, spreadsheet based, was introduced, removing double entry and improving the speed of information transfer to the production department

- Packaging area — foam-cutting machine changeovers examined and reduced significantly.

The staff in each of the key sections were heavily involved in identifying the root cause of problems in their areas. Also, there was a good degree of cross-departmental fertilisation at operator level with good effective suggestions being made and implemented. Staff wanted to be involved.

The production process was taken under control, though the production system remained as a departmentalised one. The layout is presented in Figure A2.2, along with a thread diagram of the physical product flow through the plant.

As initial successes were achieved with the WCM programme the company began to consider the move to a cellular manufacturing layout. The management team were very slow to consider any changes to the basic process and were concerned about trying new suggestions. The management team was quite young, with the General Manager in her early thirties and the rest of her team in their late twenties. The production manager decided to proceed with a number of steps that he felt could be achieved without

FIGURE A2.2: TRULIFE ORIGINAL LAYOUT

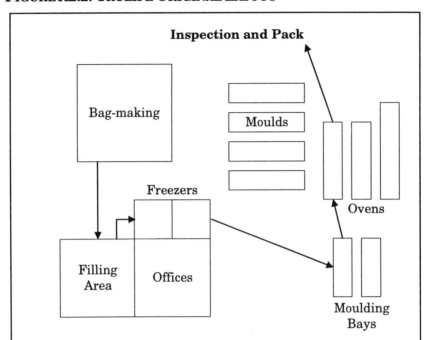

affecting the rest of the management team. The move to a cellular system of manufacture met with a high degree of resistance from one middle manager. Eventually, after much discussion, the company selected a given product family for a pilot exercise, analysed the sales for the product by size, contacted the main customer regarding a forecast of requirements and implemented a cell for the manufacture of the product. The management team felt that the customer would not co-operate with their requests and were happily surprised when they responded positively. The request was made by the marketing department and this positive response further strengthened links between the production and marketing departments.

The simplification of the manufacturing process and the subsequent high level of control of the operation meant that the management team were able to develop a simple production control system that integrated with their existing systems rather than investing in a full production control system. This saved the company nearly £70,000.

Present Situation and Plans Forward

The membership of the WCM team has changed by about 20 per cent since the start of the project. New team members have joined the team and taken over projects as their own.

The company has achieved significant operational improvements as well as improvements in the teamwork of the management team. The process of developing the company's systems is continuing, with the integration of more and more operational staff into teams to address particular problem areas. The present layout of the plant is presented in Figure A2.3. Management plan to continue their WCM efforts to gain further improvements internally.

FIGURE A2.3: TRULIFE PRESENT LAYOUT

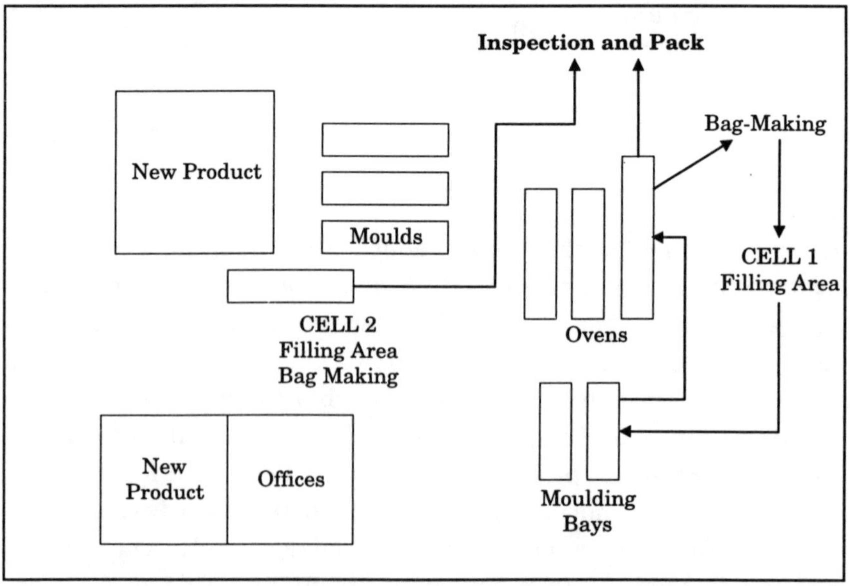

Results

The WCM programme in Trulife started in March 1994 with the diagnostic phase. During the first nine months of the programme a number of pilot projects were completed, such as the development of the quick changeover bag-making machine, and developing an understanding of the bag-making process itself.

Management developed their understanding of the basic concepts of WCM and worked through a number of initiatives in both the production and administrative areas themselves, saving, for example, approximately 50 per cent of the effort involved in getting an order entered onto the company system, removing a day from the order entry process period. The following specific savings have been identified as expressly attributable to the WCM initiative:

- Bag-making machine changeover reduced by 45 per cent

- Foam-cutting machine changeover reduced by 40 per cent

- Lead time reduced by 25 per cent

- Rejects down by 50 per cent

- Labelling reduced from 583 items to 45

- Packaging reduced by 30 per cent plus

- Instituted pilot cell. Bag-making machine has been integrated into the cell. Capacity has risen by 50 per cent with the same staff.

    ◊ Reduced WIP of bags

    ◊ Operator now available for moulding on Fridays

    ◊ Reduced quality problems on bags

    ◊ Reduced rejects as problems with bags are identified immediately and rectified.

- Packaging — alternative supplier located at a significant cost saving

- Overall units produced up by 30 per cent

- Customers have discussed forecast requirements based on Pareto Analysis and sales analysis carried out by Trulife. An agreed schedule has been put in place for deliveries. This it is believed will help production significantly as there were traditionally large swings in the order quantities placed by this major customer.

The company is actively proceeding with its WCM initiatives. Teams are being used throughout the operation to address specific problem areas. These areas have been identified as problems following a Value Added Analysis of the process involved. Simplified materials handling procedures are being implemented. Administration systems are also under close scrutiny and receiving attention on an on-going basis.

The recent implementation of a pilot cell where all the elements of manufacturing a product are brought physically together has shown dramatic and immediate improvements for the company. Staff members have co-operated fully in this change and are acting as change drivers themselves, demanding of management that they move from this pilot cell to a full implementation as soon as is practical. The cell has close-coupled workers and machines, removed lead time of approximately two days, reduced defects by nearly 50 per cent without adversely affecting the workers. On the contrary, the workers are happy to have been involved in the improvement process and they are working actively with management to further develop the system.

Index